GODS AND ROBOTS

GODS AND ROBOTS

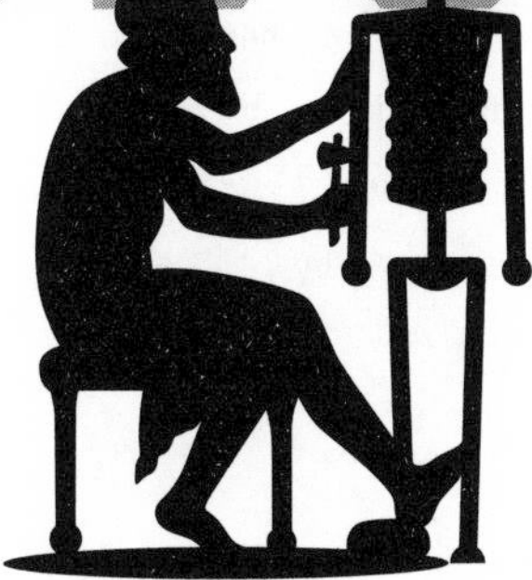

MYTHS, MACHINES, AND ANCIENT DREAMS OF TECHNOLOGY

ADRIENNE MAYOR

PRINCETON UNIVERSITY PRESS PRINCETON & OXFORD

Published by Princeton University Press
41 William Street, Princeton, New Jersey 08540
6 Oxford Street, Woodstock, Oxfordshire OX20 1TR

press.princeton.edu

Library of Congress Control Number: 2018938106
ISBN 978-0-691-18351-0

British Library Cataloging-in-Publication Data is available

Editorial: Rob Tempio and Matt Rohal
Production Editorial: Lauren Lepow
Text Design: Chris Ferrante
Jacket/Cover Design: Jason Alejandro
Production: Jacquie Poirier
Publicity: Julia Haav

This book has been composed in Adobe Text Pro, Abolition, and Refuel

Printed on acid-free paper. ∞

Printed in the United States of America

1 3 5 7 9 10 8 6 4 2

for my brother

MARK MAYOR

I sometimes wonder
whether robots were invented
to answer philosophers' questions

—TIK-TOK

CONTENTS

ILLUSTRATIONS

COLOR PLATES *(following p. 184)*

FIGURES

ACKNOWLEDGMENTS

INSPIRED IN PART by the eidetic images of the wicked robot Maria in the silent film *Metropolis* (1927) and the bronze android Talos in *Jason and the Argonauts* (1963), I started collecting ancient literary evidence for animated statues many years ago. I began to think seriously about how ancient Greek myths expressed ideas about artificial life in 2007, when I was asked to write a historical essay for the Biotechnique Exhibit catalogue, curated by Philip Ross at Yerba Buena Center for the Arts, San Francisco. My essays about Talos and Medea's experiments in rejuvenation appeared in the history of science website Wonders and Marvels in 2012. In 2016, the editors of *Aeon* invited me to write an essay about the modern relevance of classical Greek myths about *biotechne*, life by craft. I presented a preview of this book in a public lecture at the Art Institute of Chicago on March 18, 2017, "The Robot and the Witch: The Ancient Greek Quest for Artificial Life."

Many friends and colleagues read and commented on drafts of chapters at various stages. I'm especially grateful to my dear readers Marcia Ober, Michelle Maskiell, Norton Wise, and Josiah Ober for their close attention and valuable suggestions for revisions. Many others shared expertise and knowledge of ancient texts, images, ideas, and sources. My thanks to Linda Albritton, Laura Ambrosini, Theo Antikas, Ziyaad Bhorat, Larissa Bonfante, Erin Brady, Signe Cohen, John Colarusso, Sam Crow, Eric Csapo, Nick D., Armand D'Angour, Nancy de Grummond, Bob Durrett, Thalassa Farkas, Deborah Gordon, Ulf Hansson, Sam Haselby, Steven Hess, Fran Keeling, Paul Keyser, Teun Koetsier, Ingrid Krauskopf, Kenneth Lapatin, Patrick Lin, Claire Lyons, Ruel Macaraeg, Ingvar Maehle, Justin Mansfield, Richard Martin, David Meadows, Vasiliki Misailidou-Despotidou, John Oakley, Walter Penrose, David Saunders, Sage Adrienne Smith, Jeffrey Spier, Jean Turfa, Claudia Wagner, Michelle Wang, and Susan Wood. I'm grateful to Carlo Canna for his essential help in obtaining images from Italian museums and to Gabriella

Tassinari for her generous discussions of Etruscan gems. Thanks to Margaret Levi, the Berggruen Institute, and the Center for Advanced Study in the Behavioral Sciences, Stanford, for supporting my research September 2018–May 2019. Sincere gratitude is due to my excellent agents Sandy Dijkstra and Andrea Cavallaro. At Princeton University Press, I'm indebted to the anonymous readers for thoughtful critiques, to Dimitri Karetnikov for help with illustrations, to Jason Alejandro and Chris Ferrante for design, and to the nimble copyediting of Lauren Lepow. Thanks to Dave Luljak for indexing. I have benefited, as always, from the insights and enthusiasm of my editor Rob Tempio.

I'm fortunate to be able to call on my sister, Michele Angel, for amazing artistic skills and technical advice about illustrations. Barbara Mayor, my mom, is a proofreading marvel. I'm lucky to have such a wonderful brother, Mark Mayor—I know he remembers how much we enjoyed watching the movie *Jason and the Argonauts* together. Most of all, I'm forever thankful for Josh, esteemed companion of my heart and mind, and a truly good man.

INTRODUCTION

MADE, NOT BORN

WHO FIRST IMAGINED the concepts of robots, automata, human enhancements, and Artificial Intelligence? Historians tend to trace the idea of the automaton back to the medieval craftsmen who developed self-moving machines. But if we cast our nets back even further, more than two thousand years ago in fact, we will find a remarkable set of ideas and imaginings that arose in mythology, stories that envisioned ways of imitating, augmenting, and surpassing natural life by means of what might be termed *biotechne*, "life through craft." In other words, we can discover the earliest inklings of what we now call biotechnology.

Long before the clockwork contraptions of the Middle Ages and the automata of early modern Europe, and even centuries before technological innovations of the Hellenistic period made sophisticated self-moving devices feasible, *ideas* about making artificial life—and qualms about replicating nature—were explored in Greek myths. Beings that were "made, not born" appeared in tales about Jason and the Argonauts, the bronze robot Talos, the techno-witch Medea, the genius craftsman Daedalus, the fire-bringer Prometheus, and Pandora, the evil fembot created by Hephaestus, the god of invention. The myths represent the earliest expressions of the timeless impulse to create artificial life. These ancient "science fictions" show how the power of imagination allowed people, from the time of Homer to Aristotle's day, to ponder how replicas of nature might be crafted. Ideas about creating artificial life were thinkable long before technology made such enterprises possible. The myths reinforce the notion that imagination is the spirit that unites myth and science. Notably, many of the automata and mechanical devices actually designed and fabricated in Greco-Roman antiquity recapitulate myths by illustrating and/or alluding to gods and heroes.

Historians of science commonly believe that ancient myths about artificial life only describe inert matter brought alive by a god's command or magician's spell. Such tales certainly exist in many cultures' mythologies. Famous examples include Adam and Eve in the Old Testament and Pygmalion's statue of Galatea in classical Greek myth. But many of the self-moving devices and automata described in the mythical traditions of Greece and Rome—and in comparable lore of ancient India and China—differ in significant ways from things animated by magic or divine fiat. These special artificial beings were thought of as manufactured products of technology, designed and constructed from scratch using the same materials and methods that human artisans used to make tools, artworks, buildings, and statues. To be sure, the robots, replicants, and self-propelled objects described in myth are wondrous—marvelous beyond anything fashioned on earth by ordinary mortals—befitting the sublime abilities of gods and legendary inventors like Daedalus. One might consider the myths about artificial life as cultural dreams, ancient thought experiments, "what-if" scenarios set in an alternate world of possibilities, an imaginary space where technology was advanced to prodigious degrees.

The common denominator of mythic automata that took the forms of animals or androids like Talos and Pandora is that they were "made, not born." In antiquity, the great heroes, monsters, and even the immortal Olympian gods of myth were the opposite: they were all, like ordinary mortals, "born, not made." This distinction was a key concept in early Christian dogma too, with orthodox creeds affirming that Jesus was "begotten, not made." The theme arises in modern science fiction as well, as in the 2017 film *Blade Runner 2049*, whose plot turns on whether certain characters are replicants, facsimiles of real humans, or biologically conceived and born humans. Since archaic times, the difference between biological birth and manufactured origin marks the border between human and nonhuman, natural and unnatural. Indeed, in the stories of artificial life gathered here, the descriptive category *made, not born* is a crucial distinction. It separates automata described as fabricated with tools from lifeless objects that were simply enlivened by command or magic.

Two gods—the divine smith Hephaestus and the Titan Prometheus—and a pair of earthbound innovators—Medea and Daedalus—were involved in Greek, Etruscan, and Roman tales of artificial life. These four figures possess superhuman ingenuity, extraordinary creativity, technical virtuosity, and superb artistic skills. The techniques, arts, crafts, methods, and tools they employ parallel those known in real life, but the mythic inventors achieve spectacular results that exaggerate and surpass the abilities and technologies available to mere mortals in the quotidian world.

With a few exceptions, in the myths as they have survived from antiquity, the inner workings and power sources of automata are not described but left to our imagination. In effect, this nontransparency renders the divinely crafted contrivances analogous to what we call "black box" technology, machines whose interior workings are mysterious. Arthur C. Clarke's famous dictum comes to mind: the more advanced the technology, the more it seems like magic. Ironically, in modern technoculture, most people are at a loss to explain how the appliances of their daily life, from smartphones and laptops to automobiles, actually work, not to mention nuclear submarines or rockets. We know these are manufactured artifacts, designed by ingenious inventors and assembled in factories, but they might as well be magic. It is often remarked that human intelligence itself is a kind of black box. And we are now entering a new level of pervasive black box technology: machine learning soon will allow Artificial Intelligence entities to amass, select, and interpret massive sets of data to make decisions and act on their own, with no human oversight or understanding of the processes. Not only will the users of AI be in the dark, but even the makers will be ignorant of the secret workings of their own creations. In a way, we will come full circle to the earliest myths about awesome, inscrutable artificial life and *biotechne*.

Finding felicitous and apt language to describe the range of automata and nonnatural beings designated in ancient mythology as *made, not born* is daunting. The magical and the mechanical often overlap in stories of artificial life that were expressed in mythic language. Even today, historians of science and technology acknowledge that *robot*, *automaton*, *cyborg*, *android*, and the like are slippery terms with no fixed definitions. I tend to use informal, conventional understandings for *android*, *robot*, *automaton*, *puppet*, *AI*, *machine*, *cyborg*, and so on, but for clarity, technical definitions are given in the text, the endnotes, and the glossary.

This book surveys the wide range of forms of artificial life in mythology, which includes tales of quests for longevity and immortality, superhuman powers borrowed from gods and animals, as well as automata and lifelike replicants endowed with motion and mind. Although the focus is on the Mediterranean world, I have included some accounts from ancient India and China as well. Even though the examples of animated statues, self-moving objects, and simulacra of nature imagined in myths, legends, and other ancient accounts are not exactly machines, robots, or AI in the modern sense, I believe that the stories collected here are "good to think with," tracing the nascent concepts and imaginings about artificial life that preceded technological actualities.

It is important to avoid projecting modern notions of mechanics and technology onto antiquity, especially in view of the fragmentary nature of the ancient corpus about artificial life. This book is not intended to suggest direct lines of influence from myth or ancient history to modern technology, although resonances with modern science are noted. Here and there, I point out similar themes in modern mythologies of fiction, film, and popular culture, and I draw parallels to scientific history to help illuminate the natural knowledge and prescience embedded in mythic material. Along the way, the age-old stories, some very familiar and others long forgotten, raise questions of free will, slavery, the origins of evil, man's limits, and what it means to be human. As the evil robot Tik-Tok in John Sladek's 1983 science-fiction novel remarks, the very idea of an automaton leads one into "deep philosophical waters," posing questions of existence, thought, creativity, perception, and reality. In the rich trove of tales from the ancient mythic imagination, one can discern the earliest traces of the awareness that manipulating nature and replicating life might unleash a swarm of ethical and practical dilemmas, further explored in the epilogue.

So much of antiquity's literary and artistic treasure has been lost over the millennia, and much of what we have is incomplete and isolated from its original contexts. It is difficult to grasp just how much of ancient literature and art has vanished. The writings—poems, epics, treatises, histories, and other texts—that survive are but a tiny slice compared to the wealth that once existed. Thousands of artistic works have come down to us, but this is a small percentage of the millions that were created. Some art historians suggest that we have only about 1 percent of the Greek vase paintings ever made. And the modicum of literature and art that remains is often randomly preserved.

These cruel facts of loss and capricious preservation make what we do have that much more precious. They also determine one's approach and path of discovery and interpretation. In a study like this, we can analyze only what has managed to persist over millennia, as if we are following a bread-crumb trail in a deep, dark wood. And the birds have eaten most of the crumbs. Another analogy for what has perished and what survived derives from the nature of devastating wildfires cutting paths of destruction, driven by winds across a landscape of grass and trees. What remains after terrible fires is what foresters call a "mosaic effect": wide swaths of burned regions punctuated by patches of flowery meadows and copses of still-green trees. The random ravages of the millennia on Greek and Roman literature and art related to artificial life have left a patchwork dominated by blackened, empty spaces dotted here and there with vital passages and pictures from antiquity. Such a mosaic pattern necessitates a wandering path between evergreen oases, fortuitously preserved and elaborated over thousands of years. Following that path, we may to try to imagine the original cultural landscape. A similar approach, "mosaic theory," is also used by intelligence analysts to try to compose a big picture by amassing small bits of information. For this book I have gathered every text and scrap of ancient poetry, myth, history, art, and philosophy related to artificial life that I have been able to find—and enough compelling evidence emerges to suggest that people of antiquity were fascinated, even obsessed, with tales of artificially creating life and augmenting natural powers.

This is all by way of saying that readers should not expect to find a simple linear route in these chapters. Instead, like Theseus following a thread to navigate the Labyrinth designed by Daedalus—and like Daedalus's little ant making its way through a convoluted seashell to its reward of honey—we follow a meandering, backtracking, twisting thread of stories and images to try to understand how ancient cultures thought about artificial life. There is a narrative arc across the chapters, but the story lines are layered and braided, as we travel along what Artificial Intelligence futurist George Zarkadakis calls the "great river network of mythic narratives with all its tributaries, crisscrossing and circling back" to familiar characters and stories, and accumulating new insights as we go.

It may come as a relief to some, after wending our way through the vast memory palace of myth, that the final chapter turns to real, historical chronology of inventors and technological innovations in classical antiquity. This historical chapter culminates in the proliferation of self-moving devices and automata in the Hellenistic era, centered in that ultimate space of imagination and invention, Alexandria, Egypt.

Together these stories, both mythical and real, reveal the surprisingly deep roots of the quest for life that is made, not born. Let us join that quest.

CHAPTER 1

THE ROBOT AND THE WITCH

TALOS AND MEDEA

THE FIRST "ROBOT" to walk the earth—in ancient Greek mythology—was a bronze giant called Talos.

Talos was an animated statue that guarded the island of Crete, one of three wondrous gifts fashioned by Hephaestus, god of the forge and patron of invention and technology. These marvels were commissioned by Zeus, for his son, Minos, the legendary first king of Crete. The other two gifts were a golden quiver of drone-like arrows that never missed their mark and Laelaps, a golden hound that always caught its prey. The bronze automaton Talos was charged with the task of defending Crete against pirates.[1]

Talos patrolled Minos's kingdom by marching around the perimeter of the large island three times each day. As an animated metal machine in the form of a man, able to carry out complex human-like actions, Talos can be spoken of as an imagined android robot, an automaton "constructed to move on its own."[2] Designed and built by Hephaestus to repel invasions, Talos was "programmed" to spot strangers and pick up and hurl boulders to sink any foreign vessels that approached Crete's shores. Talos possessed another capability too, modeled on a human trait. In close combat, the mechanical giant could perform a ghastly perversion of the universal gesture of human warmth, the embrace. With the ability to heat his bronze body red-hot, Talos would hug victims to his chest and roast them alive.

The automaton's most memorable appearance in mythology occurs near the end of the *Argonautica*, the epic poem by Apollonius of Rhodes describing the adventures of the Greek hero Jason and the Argonauts

FIG. 1.1. Talos, bronze cast of the crumbling original model made by Ray Harryhausen for the film *Jason and the Argonauts* (1963), forged 2014 by Simon Fearnhamm, Raven Armoury, Dunmow Road, Thaxted, Essex, England.

and their quest for the Golden Fleece. Today the Talos episode is familiar to many thanks to the unforgettable stop-motion animation of the bronze robot created by Ray Harryhausen for the cult film *Jason and the Argonauts* (1963; fig. 1.1 is a bronze cast of the original model).[3]

When he composed his epic poem *Argonautica* in the third century BC, Apollonius drew on much older oral and written versions of the myths of Jason, Medea, and Talos, stories that were already well known to his audience. An antiquarian writing in a deliberately archaic style, at one point Apollonius casts Talos as a survivor or relict from the "Age of Bronze Men." This was an ornate allusion to a conceit in a figurative passage about the deep past taken from the poet Hesiod's *Works and Days* (750–650 BC).[4] In the *Argonautica* and other versions of the myth, however, Talos was described as a technological production, envisioned as a bronze automaton constructed by Hephaestus and placed on Crete to do a job. Talos's abilities were powered by an internal system of divine ichor, the "blood" of the immortal gods. This raises questions: Was Talos immortal? Was he a soulless machine or a sentient being? These uncertainties would prove crucial to the Argonauts, although the answers remain ambiguous.

• • •

In the final book of the *Argonautica*, Jason and the Argonauts are homeward bound with the precious Golden Fleece. But their ship, the *Argo*, has been becalmed. With no winds to fill their sails, exhausted from days of rowing, the Argonauts make their way into a sheltered bay between two high cliffs on Crete. Immediately Talos spots them. The great bronze warrior begins breaking off rocks from the cliff and heaves them at the ship. How could the Argonauts escape the clutches of this monstrous android? Quaking in fear, the sailors desperately attempt to flee the terrifying colossus astride the rocky harbor.

It is the sorceress Medea who comes to their rescue.

A beautiful princess from the kingdom of Colchis on the Black Sea, the land of the Golden Fleece, Medea was a bewitching femme fatale with her own set of mythic adventures. She possessed the keys to youth and age, life and death. She could hypnotize man and beast, and she could cast spells and brew powerful potions. Medea understood how to defend against flames, and she knew the secrets of the unquenchable "liquid fire" known as "Medea's oil," a reference to volatile naphtha from natural petroleum wells around the Caspian Sea. In Seneca's tragedy *Medea* (lines 820–30, written in the first century AD), the sorceress keeps this "magical fire" in an airtight golden casket and claims that the fire-bringer Prometheus himself taught her how to store its powers.[5]

Before their landfall in Crete, Medea had already helped Jason on his expedition to win the Golden Fleece. Medea's father, King Aeetes, promised to give Jason the Fleece if he could complete an impossible, deadly task. Aeetes owned a pair of hulking bronze bulls created by Hephaestus. Aeetes ordered Jason to yoke the fire-breathing bronze beasts and plow a field while sowing the earth with dragon's teeth that would sprout an instant army of android soldiers. Medea decided to save the handsome hero from certain death, and she and Jason became lovers (for the full story of how Jason dealt with the robo-bulls and the dragon-teeth army, see chapter 4).[6]

The lovers had to flee the enraged King Aeetes. Medea—whose own golden chariot was drawn by a pair of tame dragons—guided Jason to the lair of the dreadful dragon that guarded the Golden Fleece. With her shrewd psychological insight, powerful *pharmaka* (drugs), and *technai* (devices), Medea overcame the dragon.[7] Murmuring incantations, dipping into her store of exotic herbs and rare substances gathered from remote crags and meadows high in the Caucasus Mountains, Medea lulled the dragon into a deep sleep and seized the Golden Fleece for Jason. Medea and Jason absconded with the prize to the *Argo*, and she accompanied the Argonauts on their homeward voyage.

Now, facing the threat of the looming bronze automaton blocking their way, Medea takes charge again. *Wait!* she commands Jason's fearful sailors. *Talos's body may be bronze, but we don't know whether he is immortal. I think I can defeat him.*

Medea (from *medeia*, "cunning," related to *medos*, "plan, devise") prepares to destroy Talos. In the *Argonautica*, Medea uses mind control and her special knowledge of the robot's physiology. She knows that the blacksmith god Hephaestus constructed Talos with a single internal artery or tube through which ichor, the ethereal life-fluid of the gods, pulsed from his head to his feet. Talos's biomimetic "vivisystem" was sealed by a bronze nail or bolt at his ankle. Medea realizes that the robot's ankle is his point of physical vulnerability.[8]

Apollonius describes Jason and the Argonauts standing back in awe, to watch the epic duel between the powerful witch and the terrible robot. Muttering mystical words to summon malevolent spirits, gnashing her teeth with fury, Medea fixes her penetrating gaze on Talos's eyes. The witch beams a kind of baleful "telepathy" that disorients the giant. Talos stumbles as he picks up another boulder to throw. A sharp rock nicks his ankle, opening the robot's single vein. As his life force bleeds away "like melted lead," Talos sways like a great pine tree chopped at the base of its trunk. With a thunderous crash, the mighty bronze giant topples onto the beach.

It is interesting to speculate about this death scene of Talos as it was depicted in the *Argonautica*. Was the vivid image influenced by the sensational collapse of a real monumental bronze statue? Scholars have suggested that Apollonius, who spent time in Rhodes, had in mind the magnificent Colossus of Rhodes, built in 280 BC with sophisticated engineering techniques involving a complex internal structure and external bronze cladding. One of the Seven Wonders of the Ancient World, it stood about 108 feet tall, roughly the size of the Statue of Liberty in New York Harbor. Unlike the mythical Talos, who spent his days in constant motion, the immense figure of Helios ("Sun") did not have moving parts but served as a lighthouse and gateway to the island. The Colossus was demolished by a powerful earthquake during Apollonius's lifetime, in 226 BC. The massive bronze statue broke off at the knees and crashed into the sea.[9]

Other models were also at hand. Apollonius was writing in the third century BC, when an array of self-moving machines and automata were being made and displayed in Alexandria, Egypt, a lively center for engineering innovations. A native of Alexandria, Apollodorus served as head of the great library there (P. Oxy. 12.41). Apollodorus's descriptions of the automaton Talos (and a drone-like eagle, chapter 6) suggest his familiarity with Alexandria's famous automated statues and mechanical devices (chapter 9).

• • •

In older versions of the Talos story, technology and psychology are even more prominent—and ambiguous. Does his metallurgic origin make Talos completely inhuman? Notably, the question of whether Talos has agency or feelings is never fully resolved in the myths. Even though he was "made, not born," Talos seems somehow tragically human, even heroic, cut down by a ruse while carrying out his assigned duties. In the other, more complex descriptions of his downfall, Medea subdues the bronze giant with her spellbinding *pharmaka*, then uses her powers of suggestion, compelling Talos to hallucinate a nightmare vision of his own violent death. Next, Medea plays on the automaton's "emotions." In these versions, Talos is portrayed as susceptible to human fears and hopes, with a kind of volition and intelligence. Medea convinces Talos that she can make him immortal—but only by removing the bronze rivet in his ankle. Talos agrees. When this essential seal on his ankle is dislodged, the ichor flows out like molten lead, and his "life" ebbs away.

For readers today, the robot's slow demise might call to mind the iconic scene in Stanley Kubrick's *2001: A Space Odyssey* (1968). As the doomed computer HAL's memory banks fade and blink out, HAL begins to recite the story of his "birth." But HAL was *made, not born*, and his "birth" is a fiction implanted by his manufacturers, much as eidetic, emotional memories are manufactured and implanted in the replicants in the *Blade Runner* films (1982, 2017). Recent studies in human-robot interactions show that people tend to anthropomorphize robots and Artificial Intelligence if the entities "act like" humans and have a name and a personal "story." Robots are not sentient, and have no subjective feelings, yet we endow self-moving objects that mimic human behavior with emotions and the ability to suffer, and we feel pangs of empathy for them when they are damaged or destroyed. In the film *Jason and the Argonauts*, despite the expressionless face of the monolithic bronze automaton, Harryhausen's astonishing animation sequence suggests glimmers of personality and intellect in Talos. In the poignant "death" scene, as his life-fluid bleeds out, the great robot struggles to breathe and gestures helplessly at his throat while his bronze body cracks and crumbles. The modern audience feels pity for "the helpless giant and regrets that he was taken in unfairly" by Medea's trick.[10]

In the fifth century BC, Talos was featured in a Greek tragedy by Sophocles (497–406 BC).[11] Unfortunately, that play is lost, but it is easy to imagine that the fate of Talos might have evoked similar pathos in antiquity. One can appreciate how oral retellings and tragic dramas would have elicited compassion for Talos, especially since he behaved in a human-like way and his name and backstory were well known. Indeed, there is ample evidence that ancient vase painters humanized Talos in illustrations of his death.

• • •

We have only fragments of the many stories about the Cretan robot that circulated in antiquity, and some versions are lost to us. Illustrations on vases and coins help to fill out the picture, and some artistic images of Talos contain details unknown in surviving literature. The coins of the city of Phaistos, one of the three great Minoan cities of Bronze Age Crete, are an example. Phaistos commemorated King Minos's bronze guardian Talos on silver coins from about 350 to 280 BC. The coins show a menacing Talos facing forward or in profile, hurling stones. No surviving ancient source says Talos had wings or flew, but on the Phaistos coins Talos has wings. The wings could be a symbolic motif that signaled his nonhuman status or they might suggest his superhuman speed as he circled the island (this would entail traveling more than 150 miles per hour by some calculations). On the reverse of some of the Phaistos coins Talos is accompanied by the Golden Hound Laelaps, one of the three engineering marvels made by Hephaestus for King Minos. The wonder-dog has its own body of ancient folklore (chapter 7).[12]

About two centuries before Apollonius wrote the *Argonautica*, Talos appeared on red-figure Greek vase paintings of about 430 to 400 BC. The details on some of the vases show that Talos's internal "biostructure,"

FIG. 1.2. Talos hurling stones on coins of Phaistos, Crete. Left, silver stater, fourth century BC (reverse shows a bull). Theodora Wilbur Fund in memory of Zoe Wilbur, 65.1291. Right, Talos in profile, bronze coin, third century BC (reverse shows the Golden Hound). Gift of Mr. and Mrs. Cornelius C. Vermeule III, 1998.616. Photographs © 2018 Museum of Fine Arts, Boston.

the ichor-filled artery system sealed by a bolt at his ankle, was already a familiar part of the story as early as the fifth century BC. The similarities and style of the scenes suggest that the vase paintings might be miniature copies of large public wall murals painted by Polygnotus and Mikon, renowned artists of Athens in the fifth century BC. The ancient Greek travel writer Pausanias (8.11.3) tells us that Mikon painted episodes from the epic saga of Jason and the Golden Fleece in the Temple of Castor and Pollux (the Dioscuri twins were honored in the Anakeion, chapter 2).

Those murals admired by Pausanias in the second century AD are now lost, but surviving images on vases reveal how Talos was imagined in the classical era. The artists show Talos as part machine, part human, whose destruction required technology. The paintings also convey a sense of pathos in his destruction. For example, the dramatic scene on the extraordinary "Talos vase," a large wine vessel made in Athens in about 410–400 BC, shows Medea mesmerizing the large man of bronze (figs. 1.3 and 1.4, plate 1).

Cradling her bowl of drugs, Medea gazes intently as Talos swoons into the arms of Castor and Pollux. In Greek myth, the Dioscuri twins had joined the Argonauts, but no surviving stories include them in the death of Talos, so this image points to a lost tale. The Talos Painter depicts Talos with a robust metal body like that of a bronze statue; his torso looks

FIG. 1.3. "Death of Talos," The metallic robot Talos swoons into the arms of Castor and Pollux, as Medea holds a bowl of drugs and gazes malevolently. Red-figure volute krater, fifth century BC, by the Talos Painter, from Ruvo, Museo Jatta, Ruvo di Puglia, Album / Art Resource, NY.

FIG. 1.4 (PLATE 1). "Death of Talos," Ruvo vase detail. Album / Art Resource, NY.

like the realistic, heavily muscled bronze chest armor worn by Greek warriors (chapter 7, fig. 7.3). Employing the same technique used for images of warriors wearing bronze "muscle armor," the artist painted Talos's entire body yellowish-white to distinguish his bronze plating from human flesh. But despite his metallic form, Talos's posture and his face are humanized to evoke empathy. One classical scholar even detects "a teardrop . . . falling from Talos' right eye," although this line might represent metallic molding or seams, like the other reddish outlines defining the robot's anatomy.[13]

An earlier (440–430 BC) vase painting on an Attic krater found in southern Italy shows Talos as a tall bearded figure reeling off balance, again struggling against Castor and Pollux (figs. 1.5, 1.6, plate 2). This

FIG. 1.5 (PLATE 2). Medea watches as Jason uses a tool to unseal the bolt in Talos's ankle held by a small winged figure of Death, as Talos collapses into the arms of Castor and Pollux. Red-figure krater, 450–400 BC, found at Montesarchio, Italy. "Cratere raffigurante la morte di Talos," Museo Archeologico del Sannio Caudino, Montesarchio, per gentile concessione del Ministero dei Beni e delle Attività Culturali e del Turismo, fototeca del Polo Museale della Campania.

FIG 1.6. Detail of the Montesarchio krater, showing Jason using a tool to remove the bolt in Talos's ankle. Drawing by Michele Angel.

scene includes several striking details confirming the technological character of Talos's vivisystem and destruction. We see Jason kneeling next to the robot's right foot, applying a tool to the small round bolt on Talos's ankle. Leaning over Jason, Medea is holding her bowl of drugs. A small winged figure of Thanatos (Death) grasps and steadies Talos's foot. Death's stance, posed on one foot with the other bent back, appears to replicate the death throes of Talos.

A similar scene showing the use of a tool appears on an Attic vase fragment of about 400 BC found in Spina, an Etruscan port on the Adriatic Sea. Talos is again seized by Castor and Pollux. At Talos's feet, Medea holds a box on her lap and a blade in her right hand, ready to remove the nail on his ankle. Another tiny winged figure of Death points at Talos's legs, heightening the suspense of the vignette.[14]

In the Greek myth of Jason and the Argonauts, the bronze colossus was a dire obstacle to be vanquished. For King Minos of Crete, however, Talos was a boon, an early warning system and frontline defense for his strong navy. Likewise, the Etruscans, dominant in northern Italy from about 700 to 500 BC, regarded the guardian Talos as a heroic figure. Greek myths were favorite subjects for Etruscans, who imported shiploads of Attic vases decorated with familiar scenes and characters from mythology. The Etruscans often gave the Hellenic stories a local spin, however, reflected in their own artworks. Talos appears on several engraved Etruscan bronze mirrors of about 500–400 BC, when Roman power was rising as a threat to Etruria.

An Etruscan mirror in the British Museum shows Talos, identified by his Etruscan name, Chaluchasu. He is struggling with two Argonauts identified, in Etruscan-language inscriptions, as Castor and Pollux. A woman leans down to open a small box while reaching out toward Talos's lower leg (see the drawing in fig. 1.7). The scene replicates the actions of Medea in the Athenian vase paintings, but the woman is labeled "Turan," the Etruscan name for the goddess of love, Aphrodite, suggesting an alternative, unknown version of the Greek myth.

Other Etruscan bronze mirrors show a victorious Talos/Chaluchasu crushing his antagonists, perhaps reflecting his ability to roast victims by hugging them to his heated chest (fig. 1.7). Scholars conclude that a local Italian tradition glorified Talos, emphasizing the bronze robot's original purpose as the guardian of Crete's shores. The mirrors show that

FIG. 1.7. Top, Talos crushing Castor and Pollux to his chest, while a woman opens a box and reaches toward Talos's ankle. Etruscan bronze mirror, about 460 BC, drawing, 1859,0301.30. © The Trustees of the British Museum. Bottom, Talos crushing two men, Etruscan bronze mirror, 30480 Antikensammlung Staatliche Museen, Berlin, photo by Sailko (Francesko Bini), 2014.

the Etruscans considered Talos/Chaluchasu as a positive heroic figure whose "invincibility helped to overpower trespassers [and] strangers" at a time when Etruscans were facing Rome's incursions into their territory.[15]

● ● ●

How ancient is the Talos tale? That is uncertain; but, as we saw, Talos appears in art of the early fifth century BC. Stories about other animated statues and self-moving devices serving the gods on Mount Olympus are found in archaic oral traditions that were first set down in writing in about 750 BC in Homer's *Iliad*, the epic poem about the legendary Trojan War set in the Bronze Age (ca. 1150 BC).[16] In classical antiquity, it was believed that King Minos of Crete had ruled three generations before the Trojan War. Renowned for his laws and for the strong navy he built to suppress piracy, Minos was treated as a "historical" ruler by the fifth-century BC historians Herodotus (3.122) and Thucydides (1.4) and later by Diodorus Siculus (4.60.3), Plutarch (*Theseus* 16), and Pausanias (3.2.4), among others. Modern archaeologists named the Minoan civilization (3000–1100 BC) after the legendary King Minos.

Minoan-era seals from Crete depict many bizarre monsters and demons, which apparently served as guardians of cities and talismans. A bull-headed man, the Minotaur, appears on some Minoan seals. One Late Minoan seal stamp, known as the Master Impression (1450–1400 BC), is quite striking. It shows a fortified city on a hill above a rocky seashore (matching the topography of Kastelli Hill, Kydonia, modern Chania, Crete, where the seal was discovered). A gigantic faceless male figure, "unusually sturdy and strongly built," looms atop the highest point of the city. The enigmatic figure does not represent Talos of Greek myth. But if this and similar seals circulated in the Greek world in antiquity, it is possible that a scene like this—a giant seemingly guarding a Minoan city—might have influenced early oral traditions about Talos defending Crete for King Minos. That is speculation, of course, and in the absence of any literary texts the meaning of the scene on the Minoan seal remains a mystery.[17]

King Minos figured in other ancient tales of technology associated with the legendary craftsman Daedalus, whose works were sometimes conflated with those of the inventor god Hephaestus (chapters 4 and 5). In any event, it is clear that Talos, the bronze automaton of Crete, was well known in Greek poetry and artworks long before Apollonius of Rhodes wrote his *Argonautica* in the third century BC. Besides Pindar (*Pythian* 4, ca. 462 BC), Apollonius's sources for Talos are unknown, but some scholars believe that the epic traditions about the *Argo*'s voyage are even older than the Trojan War stories.[18] So the tale of Talos could be very ancient indeed.

Talos appeared in the lost tragedy *Daedalus* by Sophocles in the fifth century BC. But the earliest written description of Talos is in a fragment of a poem by Simonides (556–468 BC). Simonides calls Talos a *phylax empsychos*, an "animated guardian," made by Hephaestus. Notably, Simonides says that before taking up guard duties on Crete, the great bronze warrior had destroyed many men by crushing them in his burning embrace on Sardinia. Sardinia, the large island west of Italy, was renowned for copper, lead, and bronze metallurgy in antiquity. Sardinia had long-standing ties to Crete dating back to the Bronze Age, and the Etruscans traded and settled in Sardinia as early as the ninth century BC.[19] During the Nuragic civilization of Sardinia, which flourished from about 950 to 700 BC, smiths forged myriad bronze figures using lost-wax casting. Nuragic sculptors employed surprisingly sophisticated tools to create a phalanx of giant stone statues that stood watch on Sardinia (see also chapter 5). Ranging from 6.5 to 8 feet tall, the imposing stone figures are concentrated at Mont'e Pramo on the west coast of the island. These remarkable Nuragic statues are the earliest anthropomorphic large sculptures in the Mediterranean region, after the colossi of Egypt.

FIG. 1.8. Ancient stone giant of Mont'e Prama, Sardinia, Nuragic culture, about 900–700 BC. National Archaeological Museum, Cagliari, Sardinia.

The enigmatic giants of Sardinia have distinctive faces: large concentric discs for eyes and small slits for mouths (fig. 1.8). It's easy to see why these simple facial features are humorously likened to those of typical modern robots in popular science fiction, such as the droid C-3PO in the *Star Wars* films (1977–2017). Since 1974, archaeologists have unearthed forty-four of the great stone men at Mont'e Prama on Sardinia. The giants are believed to have served as sacred guardians. If so, they would have fulfilled the same function as Talos and other border-protecting statues in antiquity.

Was the poet Pindar's claim that the giant automaton Talos had once defended Sardinia somehow related to ancient Greek observations or reports of the towering stone giants of the island? Curiously enough, an island defended by boulder-hurling giants, the Laestrygonians, appears in Homer's *Odyssey* (10.82, 23.318). The Laestrygonians' name sounds similar to that of the Lestriconi, a tribe that inhabited northwest Sardinia. It has been suggested that the Homeric tale of the giants defending the island by throwing rocks could have arisen from sailors' sightings of the colossal figures on Sardinia.[20] The similarity to the actions of Talos is striking.

• • •

Some modern historians of automata have misunderstood Talos as inert matter supernaturally instilled with life by the gods via magic. In his history of European automata, for example, Minsoo Kang divides the automata described in antiquity into four categories: (1) mythic creatures that resemble modern robots only in appearance; (2) mythic objects of human manufacture brought to life with magic; (3) historical objects of human design; and (4) speculative automata in theoretical inquiries of moral concepts. Kang places Talos in his first category of "mythic creatures" that look like robots but were created by "supernatural power with no reference to mechanical craft." The "imaginary significance" of automata like Talos "in the premodern period had little to do with mechanistic ideas," asserts Kang, who claims that Talos was "not a mechanical being but very much a living creature."[21] But ancient sources describe Talos as "made, not born." As we saw, Talos's internal anatomy and movements were explained through mechanistic concepts, and this was echoed in ancient artistic depictions: What living creature has a metallic body and a nonblood circulatory system sealed with a bolt? Moreover, the mythic accounts and fifth-century BC artworks illustrating the destruction of Talos show that his demise required technology, specifically the removal of the bolt.

The exact definition of the term *robot* is debatable, but the basic conditions are met by Talos: a self-moving android with a power source that provides energy, "programmed" to "sense" its surroundings and possessing a kind of "intelligence" or way of processing data to "decide" to interact with the environment to perform actions or tasks. Kang's notion that ancient ideas about technology played no role in the Talos myth is based, first, on a faulty comparison to the divine creation of Adam from mud or clay in the Old Testament, and, second, on a cursory reading of the one passage in the *Argonautica* (4.1638–42) referring to Talos as the last of a "race of bronze men," the archaizing poetic trope mentioned above.[22]

Philosopher of science Sylvia Berryman maintains that the Olympian gods were not portrayed as using technology in Greek myths, and that devices made by Hephaestus were not animated by craft. But Talos's maker, Hephaestus, was the god of metallurgy, technology, and invention, usually depicted at work with his tools, and his productions were imagined as designed and constructed with implements and craft. In Berryman's view, Talos cannot represent a "technologically produced working artifact" because he has no "physical means by which [he] is said to work."[23] But Talos is outstanding among mythic artificial beings because ancient writers and artists represented Talos as an *automaton*, a "self-mover," a bronze statue animated by "an internal mechanism," in this case the single tube or vessel containing a special fluid, a system that was described in biological, medical, and machine-like terms.

Classical historian Clara Bosak-Schroeder cautions, rightly, that we moderns must guard against "projecting our technological understandings onto the past." She suggests that in similar fashion the Hellenistic Greeks might have projected their knowledge of innovations back onto their ancient myths. Following Kang and Berryman, Bosak-Schroeder assumes that all mythic examples of "automata were originally imagined as purely magical," and states that "the advent of advanced mechanics later in antiquity . . . caused Greeks in the Hellenistic and Roman ages to reinterpret magical automata as mechanical." But the argument that a form of "relative modernism" led the Greeks to retroject their current technology onto imaginary automata in their myths and legends does not apply in the case of Talos and other mythic examples of artificial life that were described as fabricated by Hesiod, Homer, Pindar, and other classical sources.[24] As discussed in chapter 9, some historical self-moving devices appeared in the fourth century BC. Moreover, Talos's features cannot be interpreted as backward projections from the Hellenistic era because, as we saw, even in the earliest versions of the myth and in artworks, Talos was already imagined as a construction, a "self-moving or self-sustaining manufactured object [that] mimicked a natural living form," the typical definition of a robot.[25]

It seems that a more meaningful, nuanced approach to Talos and other animated statues of antiquity would recognize how "mythology blurs the distinction between technology and divine power."[26] There is a difference between stories of gods wishing or commanding inert matter to become alive, as in the biblical Adam and the myth of Pygmalion's statue (chapter 6), and gods using superior forms of technology to construct artificial life, even if the inner workings are not described. As numerous scholars have pointed out, in myths about crafted beings like Talos, Pandora, and others, the artificial beings are seen as the products of divine artisanship, not just divine will. Indeed, "the mystical and technological approaches to making artificial life are not as distinct" as many believe, argues E. R. Truitt, a historian of medieval automata. Truitt explains that the promise of technologies such as metalworking "was precisely that it offered the possibility of surpassing" the ordinary limitations of human creations and ingenuity.[27]

In many of the ancient myths and legends presented in this book, artificial beings are made of the same substances and by the same methods that human craftspeople use to make tools, instruments, weapons, statues, buildings, devices, and artworks, but with marvelous results befitting divine expertise. Talos and his ilk are examples of artificial beings created, not simply by magic spells or divine fiat, as many historians and philosophers of science and technology have assumed, but by what ancient Greeks might have called *biotechne,* from *bios* "life" and *techne,* "crafted through art or science."[28]

Hephaestus, the smith god of invention, fabricated Talos in his heavenly foundry, which was imagined as resembling but far surpassing real bronze foundries on earth—with vastly superior technology, capable of producing "living" and self-moving machines (chapter 7). Bronze, an alloy of copper and tin, was the hardest, most durable man-made material of the eponymous Bronze Age. In the subsequent Iron Age, arcane bronze and bronze-making technologies retained an aura of the supernatural among ordinary folk. In popular superstition, figures made of bronze were believed to enchant or to ward off evil. Bronze guardian statues were often placed at borders, boundaries, bridges, gates, and harbors.[29] The brazen forms of the mythical Talos of Crete and the historical Colossus of Rhodes might have been thought to exert a kind of magic-shield effect, but both were engineered with complex internal structures.

From antiquity into the Middle Ages, bronze was the favored material for making "living machines" and automata. Not only did bronze casting require "trade secrets," esoteric knowledge and skills, but casting could reproduce human and animal forms in metal with a preternatural verisimilitude. This may have led to early Greek smiths being "perceived as magicians," notes Sandra Blakely in her history of metallurgy. But, Blakely continues, "to call an artisan a magician may simply be hyperbolic praise of his technical skills," especially in the case of "artifacts that seem to come alive." In the lost-wax method of bronze casting, described below, the likeness of a person or animal can seem to appear by magic. As science-fiction futurist Arthur C. Clarke's well-known Third Law states: "Any sufficiently advanced technology is indistinguishable from magic." By creating an eerie imitation of a living thing, an inventive god—or human inventor—might also "seek to replicate the animation" of that thing.[30] In the logic of magical thinking, the bronze object's uncanny replication of life suggests the notion that the simulacrum might also include self-movement and agency.[31]

Attributing magic to metallurgy could also reflect technological mastery of natural science extrapolated to metalworking, remarks Blakely. According to ancient Greek legend, the discovery of the art of pouring molten metal into crucibles occurred after a forest fire on a mountain. The "intense heat melted the ores hidden inside the earth," and as the molten ores flowed down the mountain, they filled cavities on the rocky surface, taking their exact forms.[32]

Contemplating the descriptions of Talos's biotechnology—the single vessel running from his head to his feet secured with a seal—and the way that once the seal was opened, the ichor poured out like molten lead, classical scholar A. B. Cook proposed an intriguing theory drawing on ancient metallurgy. Cook suggested that the distinctive physiology of

Talos might have symbolized or alluded to lost-wax casting in the Bronze Age. Like other bronze figurines and large bronze statues of antiquity, Talos himself would have been wrought by a lost-wax method.[33]

A finely detailed early fifth-century BC red-figure cup in Berlin, the Foundry Vase, illustrates artisans creating two lifelike bronze statues using foundry tools and techniques, including the sophisticated lost-wax method. The statue of an athlete is in process, with parts of the body as yet unconnected (fig. 1.9, plate 3; compare figs 6.3–11 for images of Prometheus constructing the first man in sections). On the other side of the vase, we see workers finishing a larger-than-life, realistic statue of a warrior (fig. 1.10).

The ancient lost-wax technology is incompletely known, but one method involved making a rough clay model or a wooden armature, which was coated with beeswax. Then the finer details were carved and molded in the wax by the sculptor. This wax model was covered with a thin clay slip, followed by successively thicker layers to make a mold. The

FIG. 1.9 (PLATE 3). Foundry scene, artisans making a realistic bronze statue of an athlete, in pieces, surrounded by blacksmith tools. Attic red-figure kylix, from Vulci, about 490–480 BC, by the Foundry Painter. Bpk Bildagentur / Photo by Johannes Laurentius / Antikensammlung, Staatliche Museen, Berlin / Art Resource, NY.

FIG. 1.10. Foundry scene, workers finishing a statue of a warrior. Attic red-figure kylix, from Vulci, about 490–480 BC, by the Foundry Painter. Bpk Bildagentur / Photo by Johannes Laurentius / Antikensammlung, Staatliche Museen, Berlin / Art Resource, NY.

core of the now-formless mass was pierced by a hollow bronze rod, from head to feet. This tube allowed the melting wax to pour out of the feet when the form was placed in a fiery furnace. Molten bronze, with lead added for plasticity and to increase flow, was next poured between the inner and outer molds where the wax had once been, to create the hollow statue. Notably, Talos heated his body by leaping into a fire, according to the poet Simonides, and his ichor flowed out at his feet.[34]

• • •

Magic and mysterious biomechanics obviously overlap in the myths about artificial life expressed in folklore terms. But in the various narratives about Talos, it is striking that the physiology of the bronze automaton was described in mytho-technical language, alluding to medical and scientific concepts current in antiquity.[35]

In the realm of myth, for example, the word *ichor* was used in a special sense for the "blood" of the gods. But in ancient medical and natural science contexts, *ichor* denoted the watery, amber-colored blood serum of mammals. Moreover, in the *Argonautica*, the poet's word for the vital vein that made up the bronze giant's circulation system was a technical term for blood vessels in Greek medical treatises. The imaginary integration of living and nonliving components, melding biology with metallurgical "mechanics," makes Talos into a kind of ancient cyborg with biomechanical body parts.[36]

Talos, as an android constructed in Hephaestus's divine foundry and animated by ichor, was presumably intended to be a perpetual-motion machine. In the myth Talos seems to evince inklings of consciousness and an "instinct" for survival, and he acquiesced to Medea's persuasion, indicating agency and volition. But Talos is unaware of his origins and does not understand his own physiology. And indeed, how should his nature be understood? According to the lost play by Sophocles, Talos was "fated to perish." And as Medea guessed, Talos was not immortal—even though ichor might have been believed to confer immortality. The myth poses a conundrum: Was Talos a kind of demigod, a "man" encased in bronze, or an animated statue?

In Greek mythology, golden ichor instead of red blood circulated in the veins of gods because they were nourished by ambrosia and nectar, which made them ageless and immortal (see chapters 3 and 4 on attempts to appropriate these divine attributes for humans). Immortal gods and goddesses could receive superficial injuries and lose a few drops of ichor without dying because their bodies quickly regenerated (Homer *Iliad* 5.364–82; cf. the fate of Prometheus, chapter 3). Even though immortal ichor flowed in Talos, Medea reasoned that if she could cause his total exsanguination, he would perish.[37]

Remarkably, the location of the robot's weak point was biologically determined. According to Hippocratic writings of 410–400 BC on bloodletting procedures, the thick vein on the ankle was the site of choice for the deliberate bleeding of patients, a traditional therapeutic operation. Writing in about 345 BC, Aristotle cited the medical writer Polybus on the major human blood vessels running from the head to the ankle, where surgeons make incisions to drain blood. One characteristic of living creatures noted by Aristotle is that their blood must remain contained in vessels as long as they live; if enough blood is lost, they swoon, but if too much is lost, they die. As early as the fifth century BC, mythographers and artists placed the nail that sealed Talos's "blood vessel" at the most logical anatomical place, corresponding to the location of the human vein known to flow most freely, so that when breached by Medea it would cause the robot to bleed out, as a human being would.[38]

The idea that Medea could destroy with the "evil eye" was an accepted notion in antiquity. According to the physical theories of some natural philosophers and other writers, certain malevolent people could send deadly rays from their eyes like psychic darts into other people, causing them harm, ill fortune, even death. Plutarch, for example, described the phenomenon as a "fiery beam" of malice emanating from an intense gaze. Medea's eyes are described as dangerous to men throughout the *Argonautica*. With her evil eye, Medea transmitted hellish phantom images (*deikela*) into Talos's being. Listening to the myth, people in antiquity would have visualized Talos's eyes as looking quite lifelike, like those of Greek bronze statues they saw: such statues were painted realistically, and their eyes were inlaid with ivory, silver, marble, and gems, with fine silver eyelashes.[39] But the evil eye should affect only living things. The idea of transmitting malevolent "rays" to disorient or destroy a machine raises the unsettling/unsettled question of Talos's true nature. A guardian made of bronze was supposed to have magical protective power. Would a metal object with no feelings be susceptible to the evil eye? That Medea could cast an evil-eye spell to disorient Talos is another indication that he was something more than an insentient metal machine.

• • •

Thousands of years before Hollywood's movie *RoboCop* (1987), about a cyborg police force, and the bionic assassins and bodyguards in the *Terminator* films (1984–2015) and other science fictions about cyborgs capable of deploying lethal force, the ancient Greeks could imagine robotic guardians created by supertechnology that imitated nature, *biotechne*. Talos, like modern ideas of cyborgs, and like other ancient automata made by divine craft, was envisioned as a hybrid of living and nonliving parts. Further, through myths like that of Talos, ancients could contemplate whether an entity "made, not born" was simply a mindless machine or an autonomous, sentient intelligence. In the Talos myth, the sorceress Medea perceived the issues that have become themes in science fiction from Mary Shelley's *Frankenstein* (1818) to *Blade Runner* (Ridley Scott, 1982) and *Blade Runner 2049* (Denis Villeneuve, 2017) to *Her* (Spike Jonze, 2013) and *Ex Machina* (Alex Garland, 2014). The Talos myth was an early exploration of the idea that automata might come to desire to be real humans. As we saw, Medea intuited that, like a mortal being, Talos might fear his own death and long for immortality.

The Talos story also showcases how the Greeks envisioned the engineering brilliance of Hephaestus, the divine smith, inventor, and technician. The myth demonstrates that at a very early date, people could conceive the idea of manufacturing a bronze android with encoded instructions to carry out complex activities based on superhuman strength: Talos could recognize and track trespassers; he could find and pick up rocks, then aim and hurl the missiles from afar. He could also crush and burn enemies within reach. Most telling, Talos could be swayed by suggestion, revealing his hybrid living/nonliving nature, the uncanny "in-betweenness" that is a persistent hallmark of automata. The Talos myth embodies age-old questions about what it is to be human and free.[40]

Some of the questions raised by the Talos tale have not escaped modern video game makers. For example, a philosophical narrative puzzle created in 2014 plumbs conundrums of Artificial Intelligence (AI), free will, and "transhumanism," the belief that advanced technology can enhance human physiology, psychology, and intelligence. The game is called *The Talos Principle*. A single player assumes the role of an AI robot that seems to have human-like consciousness and autonomy. Progressing through a complex world littered with classical ruins and relics of a lost modern dystopia, the player reacts to obstacles, clues, and choices to solve metaphysical dilemmas.[41]

More than twenty-five hundred years ago, the story of Talos set in motion ancient versions of the knotty questions about how to control automata, foreshadowing modern moral qualms that surround our robot-AI technologies. Some four hundred years ago, in 1596, poet Edmund Spenser employed a Talos-like figure—a mechanical android he named Talus—to address ethical issues of robots in *The Faerie Queene*. Can moral values be mechanized? Can machines understand justice or compassion? In Spenser's allegorical epic poem, the automated squire made of iron was sent to help Sir Artegall, the righteous cavalier, in his quest to serve justice to villains. Invincible and relentless, the Iron Knight

FIG. 1.11. Sir Artegall and his automaton squire, the Iron Knight Talus. Edmund Spenser, *The Faerie Queene* (1596), wood engraving by Agnes Miller Parker, 1953.

Talus takes his job literally. Becoming an inflexible killing machine without mercy, Talus is a symbol of an inhumane, unbending form of justice, with no interest in wrongdoers' extenuating circumstances, motives, or backstories. Concerns about whether automata can be "programmed" with ethical values (to be "artificial moral agents," AMAs, in robotic literature today), or whether automata could have emotions or "intuitions," arose in ancient and medieval myths long before sweeping advances in technology made the questions so urgent.[42]

It may seem desirable to have a security system that dispatches guardians or agents created by superior intelligence to automatically perform preordained duties triggered by specific situations. But what if the situation shifts or it becomes necessary to interrupt the automatic response? How can humans control, disable, or destroy a powerful, unstoppable machine? How does one incapacitate an automated entity once set on track?

In the ancient myth of Talos, Medea's duel with Talos turned on a twofold approach. Her knowledge of the robot's internal system allowed her to exploit a physical flaw. She also perceived that the android might have evolved human-like "emotions," such as a terror of termination. Armed with these two insights, Medea devised a trick and persuaded Talos to allow her to perform a technological-surgical operation on his body that would in fact annihilate him instead of fulfilling his innate drive or "wish" to go on forever.

The destruction of Talos was not the only time the techno-wizard Medea would wield her knowledge of artificial life to destroy an enemy by promising to cheat death.

TALOS IN THE MODERN WORLD

The solitary conduit carrying the mysterious force that animated Talos has been compared to an alternating electrical current. Bronze, being mostly copper, does have high electrical conductivity, but this fact was unknown in antiquity (although bronze colossi would have acted as lightning rods). In 2017, a writer for *Popular Mechanics* compared Talos's ichor to the blue liquid that bleeds from imaginary humanoid robots in the popular television series *HUMANS* (their animating fluid is described as a "synthetic magneto hydrodynamic conductant"). The ancient image of Talos's solitary conduit of mysterious ichor fluid may reflect something akin to what cognitive scientists call "intuitive theories" of children and adults about physics and biology. Even among people today who understand that an electrical circuit requires two wires, a mental picture persists of an empowering "juice" flowing through a single cable. Our "prescientific" intuitive vision coexists with modern scientific knowledge.[43]

In 1958, the author of a brief history of robots in *Popular Electronics* remarked on Talos's "single 'vein' running from his neck to his ankle, stoppered somewhere in his foot by a large bronze pin." Viewed in "modern terms," the author mused, this conduit "could have been his main power cable and the pin his fuse." Writing at the height of the Cold War, the author went on to declare that Talos was an ancient "Weapons Alert System and Guided Missile in one package!"[44]

Notably, that same year, 1958, the largest surface-to-air guided missile became operational. Fittingly, given Talos's role as an automated adjunct of the superior Minoan navy, the new US naval weapon system was named Talos. When development began in 1947, the military planners sought "an appropriate name." They found it in Thomas Bulfinch's popular *Age of Fable* (1855). According to the official history of the missile, Talos "watched over and guarded the island of Crete. He was made of brass and was reputed to fly through the air at such terrific speed that he became red hot. His method of dealing with his enemies was to clasp them tightly to his breast, turning them to cinders at once." In this modern telling, Talos was airborne, recalling the winged images of Talos on the coins of Phaistos, and he was heated by intense friction, but these details are not found in any Bulfinch edition or ancient text.

FIG. 1.12. Talos RIM-8 missile, 1950s. US Army/Navy archives.

Talos was "approved as the name for the new ramjet missile" in 1948. The Talos guided missiles patrolled the seas mounted on large naval carriers, ready to launch their warheads at enemies. Paralleling the duties of the mythical bronze robot on Crete, the Talos missiles served as a frontline defense, with a range of two hundred miles and a speed of Mach 2.5 (almost 2,000 mph, twelve times the estimated speed of bronze Talos). Like Talos ceaselessly circling his territory, spotting and tracking invaders, and then lobbing rocks to destroy foes, the Talos defense system was automatically directed, but it was partly autonomous at closer range. The Talos guided missiles "rode" a radar beam most of the way to the vicinity of the target but then homed in on the target "semiactively."[45]

Modern military fascination with the myth of the great bronze robot continued. In 2013, inspired by the age-old science fiction of an invincible warrior made of the strongest materials and most advanced technology, the US Special Operations Command (SOCOM) and Defense Advanced Research Projects Agency (DARPA) initiated a project to create a futuristic, robotic exoskeleton suit of armor for special operations (special ops) soldiers, something akin to the weaponized suit worn by the superhero in the film *Iron Man* (2008). Human enhancement and augmenting mortal powers are very ancient ideas, as we'll see in chapter 3. The idea for the high-tech armored suit arose from a commander's desire to protect his men in unconventional battle situations in Afghanistan and Iraq. With the Greek myth of Talos in mind, SOCOM devised the name Tactical Assault Light Operator Suit in order to render the acronym TALOS. The full-body form-fitting powered armor, intended to provide superhuman strength, hypersensory awareness, and ballistic protection, includes embedded computers, biosensors, enhanced vision and audio capabilities, solar panels, and features that capture kinetic energy. The plans for TALOS even call for an electronically activated "liquid body armor" system developed by MIT, which cannot help but recall the ichor of the immortal gods. As of this writing in 2018, TALOS is still unrealized.[46]

FIG. 1.13. TALOS, Tactical Assault Light Operator Suit, soldier's exoskeleton uniform proposal, US SOCOM.

CHAPTER 2

MEDEA'S CAULDRON OF REJUVENATION

IN THE FURTHER adventures of Jason and the Argonauts, the sorceress Medea again came to their rescue. After capturing the Golden Fleece and overcoming Talos on Crete, the Argonauts sailed home to Greece with the precious Fleece. Jason looked forward to returning to Iolcos, his hometown in Thessaly. But he found his rightful kingdom in the hands of his uncle Pelias. It was the power-mad Pelias who had commanded Jason to undertake the daunting expedition in the first place, assuming Jason would never return alive to claim the throne. Now, back in Iolcos, Jason mourned how frail his aged father, Aeson, had become.

Jason asked Medea to restore his father's youthful vigor by transferring some of his own allotted years to Aeson. But Medea rejected the notion of reducing Jason's lifespan to increase Aeson's. She chided Jason that such an exchange would be unfair, unreasonable, and disallowed by the gods. Instead she decided to try to make the old man young again through her own arcane arts.[1]

Medea's mission to revivify Aeson provides a quintessential example of mythical *biotechne* to bring about unnaturally extended life, a form of artificial human enhancement. The many different versions of this myth speculate, in folklore terms, on how one might reverse aging and increase natural life expectancy not only by casting a magical spell, but by employing certain techniques, procedures, special equipment, *pharmaka* (drugs), and therapeutic infusions.

The story of Aeson's miraculous rejuvenation by Medea's witchcraft and *pharmaka* is very old. We know that the episode was described in the *Nostoi* (*Returns*), a Greek saga based on a collection of archaic oral traditions about the aftermath of the legendary Trojan War, set in the Bronze Age. These old tales were first written down in epic form in the seventh or sixth century BC. Sadly, the full poem no longer survives. In the incomplete *Nostoi* text, however, we do learn that Medea "made Aeson a young man in his prime, stripping off his old age . . . by boiling quantities of *pharmaka* in golden cauldrons." Some ancient accounts say Medea placed Aeson himself in the kettle.[2]

According to a fragment of a lost play by Aeschylus (*Nurses of Dionysus*, fifth century BC), Medea also rejuvenated the god Dionysus's human nursemaids and their husbands by boiling them in a gold cauldron. In the fourth century BC, a contemporary of Aristotle named Palaephatus (43 *Medea*) floated a practical, if strained, "rational" explanation for the myths of Medea's rejuvenation of Aeson, Pelias, and others. Medea, he suggested, was a real woman who had discovered new, secret ways for men to seem younger. She invented invigorating steam baths created by boiling water, but the hot vapor was fatal for feeble old men. In Palaephatus's theory, it was the secrecy surrounding Medea's youth-giving therapy that led to the mythic traditions about her wondrous cauldron.[3]

At any rate, a great many writers and artists, from antiquity to modern times, retold the popular myth in dramatic imagery, depicting the witch Medea combining magical rituals with mysterious biomedical methods to reinvigorate old men.

In the literary version of the myth recounted by the poet Ovid (b. 43 BC), Medea devised the rejuvenation experiment as an audacious test of her own powers of medico-sorcery. She used a cryptic *biotechne* procedure reminiscent of her bloodletting operation on the bronze robot Talos (chapter 1). In this case, however, Medea drew all the blood from old Aeson's veins and then replaced it with a secret concoction of health-giving plant juices and other ingredients, brewed in her special vessels made of gold. Gold was recognized in antiquity to be a nontarnishing metal uncorrupted by chemical and metallic mixtures. After Medea's operation, Aeson's renewed energy and glowing vitality amazed everyone. Historians of surgery have pointed out that Medea's imaginary experiment presages modern blood transfusions, especially exchange or substitution transfusion, whereby a patient is exsanguinated and the blood replaced with donor's blood. Since 2005, for example, blood exchange experiments between young and old mice have been shown to rejuvenate the muscles and livers of the older ones.[4]

• • •

The myth of Jason and Medea in Iolcos continued with the usurper Pelias murdering members of Jason's family. In a malicious reversal of Medea's restorative blood work for Jason's old father, the evil Pelias compelled Aeson to commit suicide by drinking blood, specifically the blood of a bull or ox. In antiquity, some historical individuals—including the Athenian politician Themistocles (d. 459 BC), the Egyptian pharaoh Psammeticus (Psamtik III, d. 525 BC), and King Midas (d. ca. 676 BC)—were said to have killed themselves by drinking bull's blood.

Why bull's blood? Notably, in his treatises on anatomy written in the fourth century BC, Aristotle reported that among all animals, bull or ox blood is the quickest to congeal. Aristotle also remarked that blood flowing from the lower body of an old ox is especially dark and thick (*History of Animals* 3.19, *Parts of Animals* 2.4). It seems that the ancient myth of Aeson's demise and the historians' reports of death by drinking bull's blood expressed traditional folk knowledge of the relatively high coagulation factor of ox blood, an effect later affirmed by Aristotle. In the myth, Pelias forced Aeson to choke to death on clotted ox blood. This ancient motif has an interesting modern parallel. Bovine thrombin (blood-clotting enzyme) has been used in modern surgery since the late 1800s. It also carries risks of fatal cross-reactions in humans.[5]

• • •

After eliminating Aeson, Pelias was determined to kill Jason and his companions. The Argonauts and their allies, greatly outnumbered by Pelias's army, were thrown into uncertainty. How could they possibly avoid death and avenge the murders of Jason's father and family?

Medea stepped forward and declared that she herself would slay King Pelias for his crimes.

Success would depend upon Medea's witchcraft, her *pharmaka* of marvelous potency, a masterful sleight of hand, and her ability to convince enemies that she could really manipulate life and death in their favor. Medea's scheme would also involve bloodletting. Her plan was cunning, but it required multiple complicated steps. The ancient versions of this myth about Medea's plot to kill Pelias are also complicated. We must piece together what survives from fragments and try to reconcile ambiguities in the literary sources and various artistic illustrations. Details do not always agree, evidence that many alternative versions once circulated. But the main thread of Medea's rejuvenation of Aeson and other mythic figures provides evidence that the idea of artificially controlling normal aging and extending life by combining magic arts and medicine to enhance human physiology arose at a very early date.

• • •

Medea's murder plot relied on Pelias's belief that Medea really had turned back aging and made Jason's elderly father, Aeson, young again by means of her mysterious golden Cauldron of Rejuvenation. Medea's first step in her plot was to fill a hollow bronze statue of the goddess Artemis with drugs of diverse effects. Medea had received a cache of powerful *pharmaka* from her aunt Circe, the sorceress in Homer's *Odyssey*, and from Hekate, the goddess of black magic.[6] This venture would be another a test of her powers. Medea told Jason that she had never before used these drugs on humans.

Next, Medea disguised herself as an old priestess of Artemis, using some of her drugs to take on the appearance of a stooped, wrinkled crone. At dawn in the guise of an old hag, Medea carried the statue of Artemis into the public square of Iolcos. Pretending to be entranced, under the influence of the goddess, Medea declared that Artemis had come to bestow honor and fortune on the king. Blustering her way into the royal palace, Medea dazzled King Pelias and his daughters, convincing them that the goddess Artemis was there in person to bless Pelias "forever and ever." Medea may have used drugs and hypnosis to cause them to hallucinate an image of the goddess Artemis, or, as Christopher Faraone speculates, the portable statue may have been animated in some fashion.[7] The king and his daughters heard the old priestess cry out: *Artemis commands me to use my extraordinary powers to banish your old age and make your body young and vigorous again!*

Pelias and his daughters knew about the magical rejuvenation of Jason's father, and now the goddess seemed to promise everlasting youth for Pelias too. To prove her expertise, the old priestess called for a basin of pure water and withdrew, locking herself in a small chamber. To their

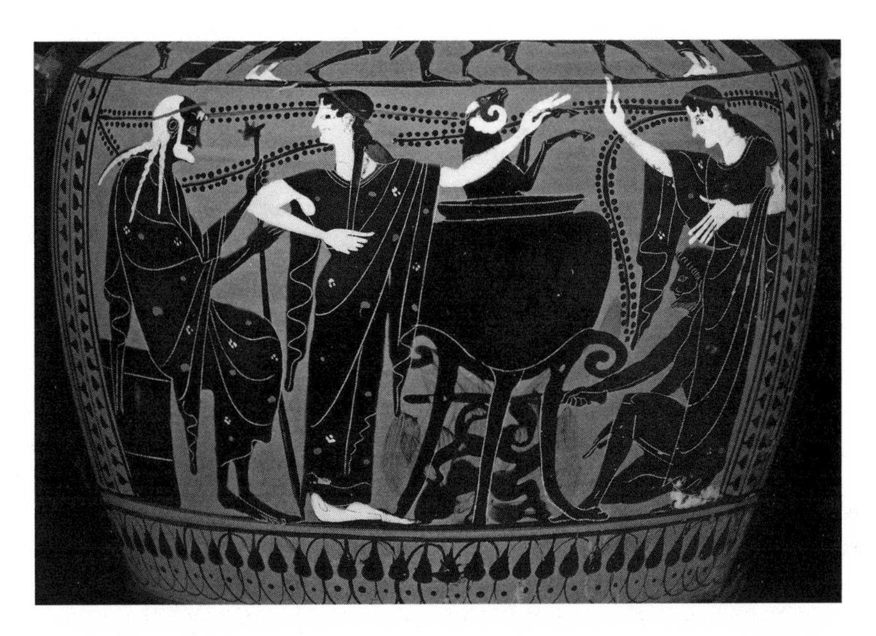

FIG. 2.1 (PLATE 5). Medea, looking back at old Pelias (left), waves her hand over the ram in the cauldron. Jason places a log on the fire, and Pelias's daughter, right, gestures in wonder. Attic black-figure hydria, Leagros Group, 510–500 BC, inv. 1843,1103.59. © The Trustees of the British Museum.

astonishment, when Medea stepped from the room, the ugly crone had transformed into a beautiful young woman. Medea promised to show Pelias's daughters how to do something similar for their elderly father.[8]

Spellbound, Pelias instructed his daughters to carry out whatever Medea commanded them to do to his body, no matter how strange it seemed. Medea invited the young women to observe a demonstration of her secret formula. They were to repeat the process exactly with their father.

In the palace, Medea recites incantations in her exotic tongue. She sprinkles the *pharmaka* from the hollow bronze statue of Artemis into her special cauldron. The daughters see Medea slit the throat of an old ram. She cuts it up and places its dismembered body in her boiling kettle. Abracadabra! A frisky young lamb magically appears!

The gullible daughters hurry away to carry out the awesome wizardry with their aged father, Pelias. Repeating the magic words, they cut their father's throat, hack up his body, and submerge him in a large pot of boiling water. Needless to say, Pelias does not emerge from the pot.[9]

Rams, lambs, and cauldrons figure in all of the literary and artistic versions of Medea's rejuvenation tales. The popularity of the motifs in Greek, Roman, and later European art shows how widespread was the fascination with the rejuvenation theme. In fifth-century BC Athens, Pelias's gruesome death at the hands of his daughters was featured in the great wall paintings illustrating Jason's adventures by the renowned artist Mikon. Mikon inscribed the names of the daughters beside their images in the Anakeion (the Temple of Castor and Pollux in Athens, Pausanias 8.11.3).

But the story of Medea's marvelous cauldron was already a very popular subject for vase painters and their customers as early as the sixth century BC.[10] Several vase paintings from about 510–500 BC show Medea bringing a ram back to life while Pelias and his daughters watch. A particularly lively example (fig. 2.1, plate 5) shows Medea waving her hand over a ram in the large cauldron. She looks back at Pelias, with a white beard and staff, who is watching intently. We see Jason placing a log under the pot, while Pelias's daughter looks on and gestures in wonder.

FIG. 2.2. Medea demonstrates the rejuvenation of a ram for Pelias, red-figure vase, about 470 BC, from Vulci, GR 1843,1103.76. © The Trustees of the British Museum / Art Resource, NY.

In a typical scene painted on a large wine jug (fifth century BC), Pelias's daughter leads him by the hand toward Medea and her cauldron with a ram inside. Another vase (470 BC, fig. 2.2) shows the ram in the kettle between Medea and Pelias. A Roman copy of a Greek marble relief of about 480–420 BC shows Pelias's daughters setting up the cauldron for Medea, who is about to open her casket of *pharmaka* (fig 2.3). The Etruscans were also fascinated by the

FIG. 2.3. Medea with the daughters of Pelias, preparing the cauldron, Roman copy of a fifth-century BC Greek marble relief, Sk 925. Bpk Bildagentur / Antikensammlung, Staatliche Museen, Berlin, photo by Jürgen Liepe / Art Resource, NY.

rejuvenation tale. A bronze mirror of the fourth century BC (fig. 2.4) shows Medea reassuringly touching the hand of an old man seated with a staff (Pelias?), while Jason places his arm around him encouragingly. A young man (Jason's old father, Aeson, rejuvenated?) emerges from a cauldron. Another woman (Pelias's daughter?) leans over Medea's shoulder, making eye contact with the old man.

FIG. 2.4. Medea and Jason reassure an old man with staff (Pelias?), as a younger man (Aeson rejuvenated?) emerges from the cauldron. Etruscan bronze mirror, fourth century BC, Cabinet de Medailles, Paris, 1329. Drawing by Michele Angel.

In an ominous scene painted about 440 BC, one daughter looks thoughtful as another helps the frail old Pelias to rise from his chair, while the third daughter waits behind a large cauldron, beckoning him and hiding a large knife by her side.[11] Yet another skillful artist painted a suspenseful scene that works as a kind of animated filmstrip around the sides of a red-figure jewelry box (fig. 2.5). Turning the box in one's hands, the viewer sees Medea carrying a sword and leading a ram toward her cauldron while Pelias's daughter beckons to her white-haired father, Pelias, who approaches Medea from the other side, leaning on his walking stick.

FIG. 2.5. The aged Pelias approaches Medea's cauldron, encouraged by his daughter. Medea beckons, while holding a sword by her side. Red-figure pyxis, late fifth century BC. Louvre. Erich Lessing / Art Resource, NY.

• • •

The ram-and-lamb motif in Medea's mytho-scientific process prefigures a modern scientific milestone involving sheep. In effect, Medea caused a young lamb to emerge from her vat of *pharmaka* mixed with the DNA of an old ram. Oddly enough, the first cloned mammal to achieve fame in popular culture was a sheep. Dolly, the genetically engineered lamb, began life in a test tube, nurtured in a growth medium "soup" in a laboratory experiment in 1996. Dolly's life ended at age six, half the life span of natural sheep and the same age as her genetic mother's cells, raising concerns that cloned animals might be destined to age and die prematurely. By 2017, scientists were able to create an artificial womb filled with man-made amniotic fluid to sustain a living lamb fetus and by 2018 they grew human cells in genetically modified sheep embryos.[12]

Cloning, genetic engineering, and artificial life-support systems have advanced apace since Dolly, of course. In the myth, Medea started with sheep and moved to human trials, paralleling the common trajectory of modern science. (The heart and lungs of sheep are about the same size and shape as human organs, which would have been noticed by the ancient Greeks.) Since 1996, many more mammal species, including primates, have been successfully cloned.

Meanwhile, the anxious ambivalence summoned by the idea of artificially meddling in the most basic natural processes of life, especially of human beings, persists. The ancient message of Medea's bold schemes to interfere with natural aging and death reverberates over the centuries. Pelias's daughters expected to recover their father's youth, as Medea's experiment appeared to promise. But they failed, horribly, to reproduce the desired results, because Medea had deliberately left out the crucial step of replacing Pelias's blood. The lurid ancient tale blurs the boundaries between charlatanism and science and deftly links the conflicting emotions of hope and horror. Hope and horror still coexist in modern Western reactions to "playing god" with science.[13]

Jason and Medea's relationship ended tragically, with Jason breaking his vows to her and Medea killing their children. Abandoning Jason, Medea escaped in her dragon chariot to other intrepid adventures. A hero but not immortal, Jason grew old and died a lonely death, crushed in his sleep by a falling timber from his rotting ship, *Argo*.

What about Medea? Was she mortal or immortal? Her ancestry might suggest that she transcended mortality. As the granddaughter of the sun god Helios and a sea nymph, Medea boasted a semidivine genealogy. In the world of myth, however, semidivine beings and demigods, nymphs, Nereids, monsters, Titans, giants, and sorceresses like Medea and Circe seem to exist in a netherworld between immortality and mortality. Medea was sometimes viewed as mortal, yet she was also portrayed as immortal and ageless. No mythic account describes her demise.[14]

In Greek myth, divinities could mate with humans, but their offspring were usually destined to perish. Medea, like many other mothers in Greek mythology, tried but failed to make her own children immortal (Pausanias 2.3.11). Yet the gods and goddesses had the power to grant everlasting life to some special humans. The Trojan boy Ganymede, for example, was abducted by Zeus's Eagle and taken up to Mount Olympus, the abode of the gods, where he remained forever young, thanks to a diet of ambrosia and nectar. And Zeus allowed the dying hero Heracles, his son by the mortal woman Alcmene, to ascend to heaven, where he was fed ambrosia, became immortal, and married Hebe, the goddess of youth (chapter 3). In another myth, Heracles's nephew, the old hero and Argonaut Iolaus, prays to Hebe and Zeus to restore his glorious youth for just one day so that he might defeat his enemy in battle. A similar tale was told about the warrior Protesilaus, who was permitted to return for one day to make love with his wife (chapter 6).[15]

Gods and goddesses never died, and they never aged either. Agelessness and immortality are closely intertwined, but they could be mutable concepts in mythology. Who besides the undying deities had ichor flowing in their veins? As we saw in chapter 1, Hephaestus gave the bronze automaton Talos ichor, but it could not guarantee his invincibility. In myth, the divine power of ichor could be transmitted to some living things too, such as plants, and even to humans, but its special effects were only temporary (see chapter 3).

In Ovid's recounting of the rejuvenation of Aeson, Medea admonishes Jason that his request to transfer years from his own life to his father was unreasonable and forbidden.[16] But Jason's request did have precedent. In the realm of myth, immortality could sometimes be shared, even traded away. For example, Heracles negotiated a bargain with Zeus to exchange the immortality of the centaur Chiron for the life of Prometheus, who was chained to a rock for stealing divine fire.[17]

And consider the confusing situation of the Dioscuri, the twins Castor and Pollux, who accompanied Jason on the *Argo* in the quest for the Golden Fleece. Mythographers could not decide whether the brothers were immortal or "half-mortal." The uncertainty arose with good reason. Their mother, Leda, was human, but Pollux was fathered by Zeus, while Castor's father was Tyndareus, a Spartan king. The novel idea of twins with different fathers posed a puzzle of mortal versus immortal bloodlines for people to ponder in antiquity. Oddly enough, the notion of twins with different paternity was not just a fantasy or plot contrivance. When two different males sire fraternal twins in the same ovulation cycle, the scientific term is *heteropaternal superfecundation*. It happens in dogs, cats, and other mammals, even including, albeit rarely, humans. Mammals can also be subject to *superfetation*, when a second ovum is fertilized while a female is already pregnant, although live human births of this kind are extremely rare because of the different rates of embryo development. The ancients were familiar with these processes, which were discussed by Herodotus (3.108) and Aristotle (*History of Animals* 585a3–9, 579b30–34), among others.[18]

In the myth of the Dioscuri, when Castor was killed, Pollux asked to share his immortality with his brother. His wish was granted by Zeus. The twins spent alternating intervals in heaven.

• • •

Behind many of the biotechno-wonders wrought by Medea, and other mythic and historical geniuses of artificial life in the coming chapters, lies a timeless theme, the search for perpetual life. Yearning to overcome death is as ancient as human consciousness. Every conscious being is born innocent of death: all human beings come into the world believing they'll live forever and be forever young. The bitter truth dawns later, a universal disillusionment that finds expression and compensation in myths around the world. The fountain of youth, the elixir of life, reincarnation, resurrection, everlasting fame in cultural memory, perpetuation of bloodlines through progeny, quests for invulnerability, grandiose building monuments—even vampires, zombies, and the undead—all testify to mortals' longing to find ways to defy death, the subject of the next chapter.

CHAPTER 3

THE QUEST FOR IMMORTALITY AND ETERNAL YOUTH

THE ANCIENT GREEKS were obsessed with eternal youth and everlasting life. In their myths, poetry, and philosophy, they devoted considerable thought to the desire to stay young and live forever. To somehow possess ageless immortality like the gods would be the ultimate achievement in a quest for artificial life. But the Greeks were also quite aware of the sobering ramifications should such boons be granted.

For the ancient Greeks, men and women's lives were measured by *chronos*, time divided into the past, present, and future. But if humans were to be set adrift in infinite time, *aeon*, what would happen to memories, or love? How might the human brain, which has evolved to accommodate seventy or eighty years' worth of memories, cope when asked to store centuries or millennia of memories? The interrelationship of human memory, love, and awareness of a finite life span was central to the modern science-fiction film *Blade Runner* (1982). The android workers in the dystopia are genetically engineered to have life spans of only four years—too short to develop a real identity based on memories or to experience empathy. In the film, renegade replicants desperately seek to increase their allotted time.[1]

The links interconnecting memory, love, and mortality also come up in Homer's *Odyssey*. In Odysseus's epic ten-year endeavor to reach his home in Ithaca after the Trojan War, he is detained against his will by the nymph Calypso. She keeps Odysseus as her lover for seven years (*Odyssey* 5.115–40). Calypso offers him eternal youth and immortality if he will stay with her on her island forever. She is incredulous when Odysseus refuses such a generous gift. The other gods insist that Calypso must honor Odysseus's desire to build a raft to try to return to his wife, family, and friends, and to live out the rest of his days in his native land. As Odysseus explains to Calypso: "I know my wife, Penelope, does not have your beauty, because she is mortal. Even so, I long to go home, despite the dangers."

Lacking empathy, the immortal Calypso cannot understand Odysseus's yearning for his wife and his nostalgia for home. As classicist Mary Lefkowitz points out, the ancient story expresses "one of the most important differences between gods and mortals. Humans have ties to each other" and to their homeland, and "the intensity of these ties is all the stronger because they cannot last." Philosopher C.D.C. Reeve suggests that Odysseus knows he will lose his identity, precious not only to him but also to his family and friends, if he chooses to become marooned in immortality.[2]

Reaching for immortality raises other profound misgivings. Unlike human individuals, immortal gods do not change or learn. "For the immortals everything is easy," notes classicist Deborah Steiner. With few exceptions, the gods act "without visible effort or strain."[3] Without the threat of danger and death, what would become of self-sacrifice, bravery, heroic striving, and glory? Like empathy, these are distinctively human ideals, and they were especially salient in a warrior culture like that of ancient Greece. The immortal gods and goddesses of Greek mythology are powerful, but no one calls the gods courageous. Undying gods, by their very nature, can never gamble on high stakes, or dare to risk obliteration, or choose to struggle heroically against insurmountable odds.[4]

● ● ●

If our lives be short—may they be glorious!

According to Herodotus (7.83), the elite infantry of ten thousand warriors in the Persian Empire of the sixth and fifth centuries BC called themselves "the Immortals," not because they wished to live forever, but because they knew that their number would always stay the same. The assurance that an equally valiant warrior would immediately take the place of each dead or wounded fighter, thereby ensuring the "immortality" of the corps, fostered a sense of cohesion and pride. The lasting appeal of the concept is evident in the name "Immortals" taken up by the Sassanid and Byzantine cavalries, by Napoleon's Imperial Guard, and by the Iranian army 1941–79.

In the great Mesopotamian epic *Gilgamesh,* the companions Enkidu and Gilgamesh face death heroically, consoling themselves that at least their fame will be everlasting. This idea is embodied in the ancient Greek ideal of *kleos aphthiton,* "imperishable glory." In Greek mythology, real heroes and heroines do not seek physical immortality. Indeed, no true hero desires to die old. Given a choice by the gods, heroic individuals like Achilles reject long lives of comfort and ease. To die young and beautiful in noble combat against an adversary who is one's match—this is the very definition of myth-worthy heroism. Even the barbarian Amazons of Greek legend achieve this vaunted heroic status, dying bravely in battle. In fact, not one ancient Amazon succumbs to old age.[5] In myth after myth, great heroes and heroines emphatically choose brief, memorable lives of honor and dignity with high-stakes risks.

That choice is the point of a legend about the Narts of the Caucasus, larger-than-life men and women who lived in the golden age of heroes. The Nart sagas combine ancient Indo-European myths and Eurasian folklore. In one saga, the Creator asks, *Do you wish to be few and live short lives but win great fame and be examples to others forevermore? Or do you prefer that your numbers be great, that you have much to eat and drink, and live long lives without ever knowing battle or glory?*

The Narts' reply is "as quick as thought itself." They choose to remain small in number and to perform bold deeds. *We do not want to be like cattle. We want to live with human dignity. If our lives are to be short, then let our fame be great!*[6]

Another antidote to wishing for immortality was the classical Greek ideal of calm, even cheerful fatalism. The attitude was plainly expressed in 454 BC, in a poem by Pindar (*Isthmian* 7.40–49) celebrating the life of a great athlete.

> Seeking whatever pleasure each day gives
> I will arrive at peaceful old age and my allotted end.

Some six hundred years later, in his *Meditations* (2 and 47) the Roman emperor and Stoic philosopher Marcus Aurelius linked the acceptance of death with one's responsibility to live one's brief, fragile life well and with honor: "Dying, too, is one of our assignments in life," he wrote. What is worthy is to "live this life out truthfully and rightfully."

• • •

Many ancient travelers' tales revel in descriptions of fabled utopias, where the people are happy, healthy, free, and long-lived. An early example of the idea that a fountain of youth or springs of longevity could be found in some exotic land of the East appears in the writings of Ctesias, a Greek physician who lived in Babylon and wrote about the wonders of India in the fifth century BC. Around the same time, Herodotus told of the long-lived Ethiopians, who owed their 120-year life span to a diet of milk and meat and their habit of bathing in violet-scented, naturally oily springs. Later, an anonymous Greek geographer living in Antioch or Alexandria (fourth century AD) wrote about the Camarini of an Eastern "Eden." They eat wild honey and pepper and live to be 120 years old. All of them know the day of their death and prepare accordingly. Curiously enough, 120 years is the maximum human life span suggested by some modern scientists.[7]

A strange little myth about an eccentric fisherman named Glaukos was the subject of a lost play by Aeschylus and a lost poem by Pindar; further details also come from Ovid, Plato, and Pausanias. In the story Glaukos noticed that when he placed the fish he caught on a special sort of grass, they revived and slithered back into the sea. Expecting to become immortal, Glaukos ate the grass and dove into the sea, where he still resides as a seer or sea daimon covered in limpets and barnacles. Another odd myth about a different Glaukos, a boy who drowned but was saved, was the subject of plays by Euripides, Sophocles, and Aeschylus (all three plays are now lost). This Glaukos was the son of King Minos of Crete. One day the little boy was playing with a ball (or a mouse) and went missing. King Minos sent the sage Polyeidus to find him. Young Glaukos was discovered dead—he had fallen into a cask of honey and drowned. But Polyeidus had once observed a snake bringing a certain plant to resurrect its dead mate. Polyeidus resuscitated the little boy with the same life-giving herb.[8]

Pliny the Elder mentioned a group of people in India who lived for millennia. India also figures in the many legends that arose after the death of

Alexander the Great, collected in the Arabic, Greek, Armenian, and other versions of the *Alexander Romance* (third century BC to sixth century AD). It was said that the young world conqueror longed for immortality. At one point, Alexander engages in philosophical dialogues with Indian sages. When he asks, "How long is it good for a man to live?" they reply, "As long as he does not regard death as better than life." In his travels, Alexander is constantly thwarted in his search for the water of everlasting life, and he meets fantastic angels and sages who warn him against such a quest. The dream of finding magic waters of immortality persisted in medieval European folklore. The legendary traveler-storyteller Prester John, for example, claimed that bathing in the fountain of youth would return one to the ideal age of thirty-two—and that one could repeat the rejuvenation as often as one liked.[9]

• • •

On the other side of the world, in China, ancient folktales told of Neverdie Land (*Pu-szu chih kuo*) where people ate a miraculous fruit.[10] Several historical emperors dreamed of discovering the elixir of immortality. The most famous seeker was Qin Shi Huang, born in 259 BC, about a century after Alexander the Great. The Taoist legends told of *ti hsien*, people who never aged or died because they cultivated a special herb on legendary mountains or islands. In 219 BC, Qin Shi Huang dispatched an alchemist and three thousand young people to try to discover the elixir. They were never seen again.

The emperor sought out magicians and other alchemists, who compounded various broths containing ingredients believed to artificially confer longevity, from hundred-year-old tortoise shells to heavy metals, especially *tan sha*, red sand or cinnabar (mercuric sulphide). In antiquity, mercury's mysterious liquid state and astonishing mobility led people to consider quicksilver a "living metal" (see chapter 5 for mercury used to power automata). Qin Shi Huang died at the relatively advanced age of forty-nine in 210 BC. His immortality came in the form of his lasting legacy as the first emperor of unified China: he was the builder of the first Great Wall, the great Lingqu Canal, a magnificent mausoleum guarded by six thousand terra-cotta warriors, and a tomb with underground rivers of mercury.[11]

In contrast to Qin Shi Huang's anxieties about dying, Marcus Aurelius (*Meditations* 47 and 74) crystallized the Stoic view, pointing out that "Alexander the Great and his mule driver both died and the same thing happened to both. They were absorbed alike into the life force of the world or dissolved alike into atoms." Think of every person and creature who has ever lived and died, "all underground for a long time now. What harm does it do them?" The historical Alexander's own acceptance of his mortality was neatly distilled in a famous quip. It was recorded by several of his biographers near the end of the arduous campaigns in India. Alexander had already conquered the Persian Empire and had survived numerous serious battle wounds. Some men in his entourage had even begun to hail him as a god. In the midst of the heavy fighting in 326 BC, an arrow pierced Alexander's ankle. As his companions rushed to his side, Alexander smiled ironically and quoted a well-known passage from Homer: "What you see here, my friends, is blood—not *ichor which flows from the wounds of the blessed immortals*."[12]

Like Alexander—who would perish young and beautiful three years later (323 BC)—the great heroes of classical antiquity ultimately came to terms with their impending physical death, consoled by winning an everlasting "life" in human memory—even though it meant they must join Homer's sad "twittering ghosts" in the Underworld.[13] The ancient myths about immortality deliver an existential message: not only is death inescapable, but human dignity, freedom, and heroism are somehow intertwined with mortality.

• • •

The flaws inherent in seeking immortality come to light in myths about the most fearless mortal heroes. Take the case of Achilles. When he was born, his mother, the Nereid Thetis, sought to make him invulnerable by anointing his body with divine ambrosia and then "burned away his mortality" by holding him over a fire. According to the more famous version of the myth, she dipped baby Achilles in the River Styx to render him immortal. In both myths, Thetis had to hold Achilles by the heel, which remained his vulnerable spot (Apollonius *Argonautica* 4.869–79; Statius *Achilleid*). Years later, on the battlefield at Troy—despite his valor—the best Greek champion did not expire in the honorable face-to-face combat that he hoped for. Achilles died ignominiously because an arrow shot by an unseen archer homed in on his heel, the seemingly insignificant weak link in his body. Likewise, the god Hephaestus and King Minos of Crete did not anticipate that the bronze robot Talos could be toppled by Medea's simple operation on his ankle that drained him of ichor (chapter 1). Unforeseen vulnerabilities are always the Achilles's heels of cutting-edge *biotechne*.

Many ancient myths also ask whether immortality can guarantee freedom from suffering and grief. For example, in the Mesopotamian epic the hero Gilgamesh resents that only the gods live forever, and he fears his own death. He sets off on a quest for the Plant of Immortality.[14] But if Gilgamesh were to achieve his desire for everlasting life, he would eternally mourn the loss of his dear mortal companion, Enkidu.

And consider the fate of the wise centaur Chiron, teacher and friend of the Greek hero Heracles. During a battle, it happened that Chiron was accidentally struck by one of Heracles's poison arrows. The arrow, tipped with venom from the Hydra monster, inflicted a terrible wound that would never heal. Wracked with unbearable pain, the centaur begged the gods to trade his immortality for blessed death. Some myths claimed that Prometheus, the Titan who secretly taught humans the divine secret of fire, offered to exchange places with Chiron. Zeus's notorious punishment of Prometheus was designed to cause interminable torture. Zeus chained Prometheus to a mountain and dispatched his Eagle to peck out his liver every day. The regenerative power of the liver was known in antiquity.[15] Accordingly, in the myth the immortal Titan's liver grew back overnight, for the Eagle devour again. And again. Forever.

A horror of monstrous regeneration also drives the myth of the many-headed Hydra monster. Struggling to kill the writhing serpent, Heracles lopped off each head, and watched aghast as two more grew back in its place. Finally he hit on the technique of cauterizing each neck with a flaming torch. But he could never destroy the immortal central head of the Hydra. Heracles buried the indestructible head in the ground and rolled a huge boulder over the spot to warn off humans. Even buried deep in the earth, however, the Hydra's fangs continue to ooze deadly venom. The myth makes the Hydra a perfect symbol of the infinitely proliferating consequences of immortality. Indeed, Heracles himself was doomed by his own Hydra-poison *biotechne*. Because he treated his arrowheads with the monster's venom, he possessed an unlimited supply of poison projectiles with their own chain of unintended disasters. The centaur Chiron was only one of the victims. The great Heracles himself perished ingloriously, in agony from secondhand Hydra venom.[16]

An interesting variation on the theme of nightmarish regeneration appears in the old story of an automaton in the form of a broom. The "Sorcerer's Apprentice" tale was recounted by Goethe in 1797 and popularly retold in the episode starring Mickey Mouse in Disney's 1940 animated film *Fantasia*. In fact, the original tale first appeared in written form in about AD 150, told by Lucian of Samosata, a novelist of satire and speculative fiction (now called science fiction).[17] In his story *Philopseudes* (*Lover of Lies*), a young Greek student travels with an Egyptian sage, a sorcerer who has the power to make household implements, such as a broom or pestle, into android servants that automatically do his bidding. One night while the sage is away, the student attempts to control the wooden pestle by himself. He dresses it in clothes and commands it to bring water. But then he cannot make the automaton stop carrying buckets of water. The inn is flooding, because he lacks the knowledge to turn the automaton back into a pestle. In desperation, the student chops the unstoppable servant with an axe, but each piece becomes another water-carrying servant. Luckily, the sage returns in time to save the day.

• • •

Several ancient Greek myths caution that cheating death causes chaos on earth and involves grievous suffering. "Sisyphean task" is a cliché connoting futile, impossible work—but few recall *why* Sisyphus must push a boulder to the top of a hill forever. Sisyphus, the legendary tyrant of Corinth, was known for his cruelty, craftiness, and deceit. According to the myth, he slyly captured and bound up Thanatos (Death) with chains. Now no living things on earth could die. Not only did this deed overturn the natural order and threaten overpopulation, but no one could sacrifice animals to the gods or eat any meat. What would happen to politics and society if tyrants lived forever? Moreover, men and women who were old and sick or wounded were condemned to suffer interminably. The war god Ares was especially irritated because if no one was in danger of dying, warfare was no longer a serious enterprise. In one version of the myth,

Ares freed Thanatos and delivered Sisyphus into the arms of Death. But then, once in the Underworld, the cunning Sisyphus managed to convince the gods to release him to rejoin the living, temporarily, to attend to some unfinished business. Thus he slipped out of Death's grasp again. In the end, Sisyphus died of old age, but he was never enrolled among the shades of the dead fluttering uselessly about the Underworld. Instead, he spends eternity in hard labor. The story of Sisyphus was the theme of tragedies by Aeschylus, Sophocles, and Euripides.[18]

In the realm of myth, then, immortality posed dilemmas for gods and humans alike. In chapter 2, the old men Aeson and Pelias sought to turn back the clock but died anyway, and the myths of Talos, Achilles, Heracles, and others also point to the impossibility of preparing for every potential design flaw in the quest to become something more than human. Yet the dream of eternal, ageless life persists.

● ● ●

The myth of Eos and Tithonus is a dramatic illustration of the jinxes that lurk in the desire to surpass a natural human life span. The tale of Tithonus is quite old, first recounted in the *Homeric Hymns*, a set of thirty-three poems mostly composed in the seventh and sixth centuries BC. The story tells how Eos (Dawn or Aurora, the "rosy-fingered" goddess of morning light) fell in love with the handsome young singer-musician of Troy named Tithonus. Eos took Tithonus to her celestial bower at the end of the earth to be her lover.

Unable to accept the inevitable death of her mortal lover, Eos fervently requested life everlasting for Tithonus. In some versions, it is Tithonus himself who longed to be immortal. At any rate, the gods granted the wish.

In typical fairy-tale logic, however, the devil was in the details. Eos had forgotten to specify eternal youth for her beloved. For him, the years pass in real time. When loathsome old age begins to weigh upon Tithonus, Eos despairs. In sorrow, she places her aged lover in a chamber behind golden doors where he remains for eternity. There, devoid of memory or even the strength to move, Tithonus babbles on endlessly. In some versions, Tithonus shrivels into a cicada, whose monotonous song is a never-ending plea for death.[19]

FIG. 3.1. Eos (Dawn) pursuing Tithonus, Attic red-figure cup, Penthesilea Painter, 470–460 BC, inv. 1836,0224.82. © The Trustees of the British Museum.

Gods and goddesses, forever young and glamorous, were believed to grieve over the death of their children conceived with mortals. In the myth, Eos and Tithonus had a son, Memnon. The Ethiopian ally of the Trojans in the legendary Trojan War, Memnon fought courageously against the Greek hero Achilles. Memnon was killed. The dewdrops that appear at dawn were said to be the tears of Eos, mourning for her son. Zeus took pity on Eos and granted her plea that Memnon would live eternally on Mount Olympus. This time, Eos remembered to request that her son would remain as young as he was at the moment of his death.[20]

Just as mortals regret their own mortality, the gods regret the mortality of their human favorites. But gods are especially averse to the natural progression of old age and decrepitude, particularly in their human lovers. In Homer's *Odyssey*, mentioned above, the nymph Calypso complained bitterly that the other gods begrudged happiness to goddesses like her and Eos who fall in love with mortal men. In the archaic *Homeric*

FIG. 3.2. Eos (Tesan) and Tithonus (Tinthun), Etruscan bronze mirror, fourth century BC, inv. 1949,0714.1. © The Trustees of the British Museum.

Hymn to Aphrodite, the goddess of love herself callously takes leave of her own mortal lover Anchises. "I would not choose to have you be immortal and suffer the fate of Tithonus," Aphrodite explains to Anchises. "If only you could retain your present appearance and stature, then we could remain together. But soon savage old age will overtake you—ruthless old age, which we gods despise as so dreadful, so wearying."[21]

Itself ageless, the Tithonus myth has been immortalized by artists and poets over millennia. Early modern artists tend to emphasize the contrast between the white-haired oldster and the ever-rosy Dawn.[22] But the myth's darker message is the focus in the ancient Greek illustrations. Vase painters depicted the young musician nervously fleeing capture by the lustful Eos, as though he already senses how the story must end. Love matches between pitiless gods and mere mortals end tragically. A similar foreboding affected the young maiden Marpessa, who was wooed by the handsome god Apollo and by a mortal named Idas. In that myth,

FIG. 3.3. Tithonus turning into a cicada, engraving, Michel de Marolles, *Tableaux du Temple des Muses* (Paris, 1655). HIP / Art Resource, NY.

Idas and Apollo fought for her hand, but Zeus allowed the girl to chose between the suitors. Marpessa chose Idas because she knew that Apollo would desert her in old age (Apollodorus *Library* 1.7.8).

A fragment of a verse by the great poet Sappho (ca. 630–570 BC) written on scraps of papyrus was deciphered in 2004. The verse is known as the Tithonus or old age poem. Lamenting that she is growing old and gray, Sappho recalls the myth of Tithonus and urges younger songstresses to revel in their music while they may. Along similar lines, in the first century BC, the Roman poet Horace refers to the misery of Tithonus and other would-be immortals in his ode (1.28) warning of the perils and the false allure of immortality, which "entails a fate worse than death." Many centuries later, in a poem penned in 1859, Alfred Lord Tennyson imagined the heartbroken Tithonus, consumed by the cruel curse of immortality, not only exiled by his unnatural longevity from his beloved's embrace but cast out of humanity. A senescent Tithonus, a pitiful shadow of a man isolated by dementia, is attended by young Eos in a haunting poem by Alicia E. Stallings ("Tithonus," *Archaic Smile*, 1999). This depressing myth about the "horror of aging" would have been forgotten thousands of years ago if the message did not somehow give people subconscious comfort about the inevitability of death, declares Aubrey de Grey, a gerontologist who seeks limitless rejuvenation through futuristic science.[23]

• • •

In the Homeric imagination, gods and goddesses remained youthful and vital forever because of their special diet. They were sustained by ambrosia and nectar, which produced ethereal ichor instead of blood. Ambrosia (the term derives from a Sanskrit word for "undying") was also a protective and rejuvenating body lotion used by goddesses (Homer *Iliad* 14.170). In the *Odyssey* (18.191–96), Aphrodite gives Odysseus's wife, Penelope, "immortal gifts" including ambrosia to maintain her youthful beauty. As with the mysterious "waters of life," the actual composition of ambrosia and nectar was never specified. Deities could give ambrosia to mortals to make them invulnerable, as Thetis attempted with her son, Achilles (above) or to confer agelessness and/or immortality on chosen humans, as was done for Heracles (chapter 2). An intriguing fragment of a poem by Ibycus (sixth century BC), preserved by Aelian (*On Animals* 6.51), refers to an ancient story about Zeus rewarding the humans who tattled on Prometheus "with a drug to ward off old age." About a thousand years later, the poet Nonnus (*Dionysiaca* 7.7) cynically complained that Prometheus should have stolen the nectar of the gods instead of fire.

Tantalus was another figure who was eternally punished for misdeeds against the gods. One of his crimes was his attempt to steal divine ambrosia and nectar to give to humans to make them immortal (Pindar *Olympian* 1.50). It is interesting that the mythic key to eternal youth and life was nutrition: the gods had a special diet of life-giving food and drink. Notably, nutrition is the most basic common denominator that distinguished living from nonliving things in Aristotle's biological system. Hoping to unravel the mysteries of longevity, Aristotle investigated aging, senescence, decay, and death in his treatises *Youth and Old Age, Life and Death,* and *Short and Long Lifespans.* Aristotle's scientific theories about aging concluded that senescence is controlled by reproduction, regeneration, and diet. The philosopher noted that sterile or continent creatures live longer than those that drain energy in sexual activity. Perhaps it is no surprise that modern life-extension researchers also focus on nutrition and caloric restriction. And Aristotle would be gratified to learn that there is indeed an evolutionary trade-off between longevity and reproduction, and that long-term modern studies suggest that sexual abstinence can add years to individuals' life spans.[24]

• • •

In all the iterations of the Tithonus myth, ancient and modern, the final image of the once-vital singer is one of lost dignity. His awful fate—"life detested but death denied"—casts a heavy shadow over the practical and spiritual problems of stretching human life spans far beyond natural limits, thanks to advances in medicine.[25] As Sophocles remarked in his play *Electra*, "Death is a debt all of us must pay." Echoing the prescience of Greek mythology, more than two millennia ago the philosopher Plato had Socrates argue that it is wrong to keep people alive when they can no longer function. Medicine, Socrates asserts, should be used only to treat curable diseases and to heal wounds, not to prolong a person's life beyond its proper time (*Republic* 405a–409e). Today, however, rejuvenation researchers and optimistic transhumanists believe that science can make death optional. Modern immortalists look forward to living indefinitely through utopian diets, medicine, and advanced *biotechne*, merging humans and machines or uploading brains into the Cloud (and its technological progeny).[26]

But human cells are naturally programmed to age and expire; bodies have evolved to be disposable vessels for transmitting genes from one generation to the next. This fact is recognized by scientists as the "Tithonus dilemma," namely, the consequences of longevity without health and vigor. The dilemma plagues the project of keeping people alive indefinitely without their bodies and brains succumbing to age and cellular decay, like Eos's tragic lover in the myth. Aubrey de Grey believes that modern humans need to overcome what he calls the "Tithonus error," the humble acquiescence to aging and death. To counter the Tithonus dilemma, he founded SENS (Strategies for Engineered Negligible Senescence) Research Foundation in 2009, with the mission of supporting scientific innovations to bypass or switch off the natural decrepitude of cells as death is increasingly postponed. Failure raises the specter of a future dystopia populated by myriad transhuman Tithonus-like wraiths, a prospect even more hellish than the Homeric Underworld of gibbering ghosts.[27]

• • •

Tithonus embodies a stark tale: for human beings, excessive life, inappropriate or unseemly survival—living too long—could be more horrifying and tragic than dying too soon. Living forever robs memories of human meaning, just as surely as a life cut too short precludes a store of memories. The Tithonus story and similar myths give voice to anxieties about "overliving," continuing to exist beyond what should mark a natural death. As we saw, overliving also concerned ancient philosophers. Those who overlive become superannuated, obsolete, pitiable. Even agelessness—eternal youth—offers no solace. This idea suffuses Anne Rice's influential modern gothic novels *The Vampire Chronicles* (1976–2016) and the film *Only Lovers Left Alive* (2013, Jim Jarmusch). The immortal, ever-youthful vampires are lost, wandering souls who grow more world-weary, more jaded and bored with each passing millennium.[28]

Overliving, overreaching: a host of myths and legends reveal the folly of seeking immortality. But if turning back old age and postponing natural death were unreasonable and forbidden, as Medea cautioned Jason (chapter 2), then could mortals at least hope to somehow enhance their physical capabilities—which are so paltry compared to those of the gods? Even some unthinking animals enjoy more magnificent powers than do weak, vulnerable human beings. Another thought-provoking body of Greek myths about artificial life investigates whether *biotechne* might be used to "upgrade" nature and somehow engineer hyperhuman powers.

CHAPTER 4

BEYOND NATURE

ENHANCED POWERS BORROWED FROM GODS AND ANIMALS

HOW DID HUMANS come to be weaker and more vulnerable than wild beasts? As Plato recounted the story, human beings were stinted because it was left to a committee of two to distribute the abilities of earthly creatures (*Protagoras* 320c–322b). After the creation of humans and animals, the gods put two Titans, Prometheus and his younger brother Epimetheus, in charge of allocating capabilities. Epimetheus ("Afterthought") was not as wise as his brother Prometheus ("Forethought"). Epimetheus begged to have the privilege of assigning various powers, promising that Prometheus could then inspect his work.

Epimetheus began sorting out the natures of animals of land, sea, and sky. He was so absorbed in the task of ensuring their survival, with gifts of speed, strength, agility, camouflage, fur, feathers, scales, keen eyesight and hearing, superb sense of smell, wings, fangs, venoms, talons, hooves, and horns, that he absentmindedly used up all the abilities on nonreasoning creatures. With a start, he realized that there was nothing left for the naked, defenseless humans, just as his brother Prometheus arrived to inspect the creatures—and on the very day they were destined to emerge on earth.[1]

"Desperate to find some means of survival for the human race," Prometheus stole the powers of technical skills, speech, and fire from the gods to bestow on the weak mortals, so that the men and women could at least make tools and figure out how to compensate for their pitiful capabilities. As Brett Rogers and Benjamin Stevens point out in their comparative study of classical Greco-Roman literature and modern science fiction, the myth of Prometheus can be read as an early "explanatory account and as a symbol for the ongoing human relationship to technology," an example of "speculative fiction" conceived by an ancient culture not usually seen as "techno-scientific." The gifts bestowed by Prometheus represent the first "human enhancements," defined as "attempts to temporarily or permanently overcome limitations of the human body by natural or artificial means."[2]

As the Greek myth tells us, Zeus sentenced Prometheus to perpetual pain, commanding his Eagle to devour the Titan's liver every day. But the Titan's gifts to humanity keep on giving, with potential for both positive and worrisome ramifications. "Technology makes up for our absurd frailty," comments Patrick Lin, a philosopher who studies the ethics of robotics, AI, and human enhancement technologies (HET). "We naked apes couldn't survive at all if it weren't for our tool-making intellect and resourcefulness." Today, human enhancements such as visual and hearing aids, titanium joints, pacemakers, stimulants, and bionic prosthetics are commonplace and welcomed.[3] But controversies arise over some human improvements and supernatural enhancements slated for questionable uses. People start to worry when, for example, military scientists seek to make soldiers "more than human" through drugs, implants, exoskeletons like the TALOS project (chapter 1), human-machine hybrids, neurorobotics, and by replicating the enviable powers of animals. As Lin and his colleagues warn, multiple practical and moral risks swarm around modern attempts to "upgrade" the bodies of humans and to develop augmented soldiers, military androids, cyborg creatures, drones, and robot-AI auxiliaries.[4] By now, it will come as no surprise that the outlines of some of those quandaries were foreshadowed in ancient Greek times.

Techne combined with intellect and audacity—these are the unique gifts that human beings rely on to survive in the world. This ancient Greek understanding was beautifully summarized by the playwright Sophocles (*Antigone* 332–71). "Humans are formidable," declared Sophocles, for no other creatures have the skills and daring to navigate the stormy seas, plow the earth, tame horses and oxen, hunt and fish, devise laws and make war, and build and rule cities; no other creatures have the facilities of language and "wind-swift thought" of "all-resourceful" humans, ceaselessly contriving ways to escape the forces of nature. "Skillful beyond hope is the contrivance of humans' inventive arts (*mechanoen technas*) which advances them sometimes to evil and other times to good."[5]

In the myths about Medea, Jason, and the legendary inventor Daedalus we find the earliest records of how humans desired to exceed and augment human powers, to create unnatural forms of life, and to harness artificial beings—including animal replicas. As we have seen, Prometheus suffers eternal punishment for giving mortals tools and fire, and Tantalus pays forever for stealing ambrosia for humans. Now, let us take a look at another myth of human enhancement, in which the cunning wizard Medea manages to make off with a quantity of divine ichor, to help Jason defend himself against superior deadly forces.

• • •

In the ongoing escapades of the Argonauts, Medea mixes a potion and devises a clever tactic to protect Jason from her father's fire-breathing brazen bulls and an army of unnatural soldiers that arise from dragon's teeth. In search of ultrapowerful *pharmaka* for her lover, Medea treks to the high Caucasus Mountains, to the rocky crag where Zeus had chained Prometheus. Medea knows that a rare flowering plant grows in the soil wherever precious ichor drips from Prometheus's side as the Eagle ravages him. When they are cut, the strange plant's flesh-like roots ooze a black sap containing the essence of the immortal Titan's ichor. Medea collects the sap in a pure white shell from the Caspian Sea and compounds a potent drug. Known as "Promethean," the ointment imparts superhuman powers, deflects fire, and resists enemy spears. The effects of the ichor-drug are spectacular, but temporary, lasting only one day.[6]

In the *Argonautica*, the Promethean ichor preparation gives the normally passive Jason incredible Herculean strength and courage. As Medea promised, Jason suddenly feels "unbounded valor and great might like that of the immortal gods." As the drug begins to circulate, he senses "terrifying powers entering his body." His arms begin to twitch and flex, his hands clenching at his sides. Like a warhorse eager for battle, Jason "exults in the superhuman strength of his limbs." Under the influence of the ichor coursing through his body, Jason "strides and leaps about, brandishing his spear and roaring like a wild beast."[7]

FIG. 4.1. Prometheus bleeding ichor on the ground, as Zeus's Eagle pecks his liver. Laconian cup, sixth century BC. Vatican Museum. Album / Art Resource, NY.

The effects of the drug as described in the *Argonautica* put one in mind of synthetic psychoactive stimulants: for example, modern street drugs chemically related to but much stronger than cathinone from *qat* plants which can cause users to feel that they have superhuman strength and goad them into ferocious acts. Today's military pharmacologists are creating "human enhancement" concoctions that could supercharge soldiers mentally and physically, making them behave much like Jason under the influence of the Promethean ichor. Millennia ago in Homer's *Odyssey* (4.219–21), Helen of Troy mixed an elixir called nepenthe, imagined as opium and wine, to dispel the traumatic memories, "anger, and grief" of the shell-shocked veterans of the Trojan War. Now military scientists seek drugs and other neurotechnological brain interventions that would allow troops to go without sleep, sense no physical pain, exceed normal aggression, override moral qualms about killing or torture, erase negative thoughts, and obliterate memories of wartime violence or atrocities.[8]

• • •

Returning to the myth of the Golden Fleece, we witness how Medea's Promethean drug lends Jason the physical and mental power to wrangle the pair of bronze robo-bulls that were forged for King Aeetes by the smith god Hephaestus. Aeetes commanded Jason to plow a field with these fire-breathing bulls, plant a helmet-full of dragon's teeth, and then defeat the invincible army that would arise from these sown dragon "seeds," all before sunset. The king is confident that even if Jason somehow manages to avoid being burned to death and plants the teeth, he and his men will be destroyed by the unstoppable automaton warriors that will spring up from the field.

At dawn, the fearsome bulls emerge from their sooty underground stalls, pawing the ground with their brazen hooves. They charge at Jason, flames shooting from their nostrils "as though blasted by bellows from a bronze-smith's furnace." Jason braves the searing breath of the oxen and yokes them to the bronze plow. All day he plows the large field and sows the dragon's teeth.[9]

It is nearly dusk when the plowed furrows begin to seethe and gleam as the "earthborn" warriors in armor sprout from the field. This is the horrid crop of robot-like soldiers that must be "harvested," cut down, before nightfall. The scene of skeleton soldiers popping out of the ground is beloved by aficionados of science fiction and classical mythology film, as it was realized in the spectacular Harryhausen sequence in *Jason and the Argonauts* (1963).

In the *Argonautica*, the "earthborn" warriors are ghostly giants clad in bronze armor, springing up fully armed, ready to attack. Luckily, Medea has instructed Jason how to deal with the multiplying, uncontrollable mob. The earthborn soldiers lack one crucial attribute: they cannot be ordered or led, nor can they retreat. They are hardwired to advance and attack. With continuous reinforcements swelling their ranks, the armed androids march on the nearest "enemy"—Jason's men.

Just as Medea figured out how to incapacitate the bronze robot Talos of Crete by exploiting his internal mechanical weakness and "almost human" artificial intelligence, she now takes advantage of the coding imprinted in the sown army. Medea advises Jason to toss a stone to trigger the soldiers' programming. She realizes that a random impact will initiate a domino effect, a cascade of blows, causing each android to fight the nearest soldier and thereby destroy each other.

As the first ranks of the dreadful army begin to advance toward the Argonauts, Jason throws a boulder into their midst. Sensing the blows striking their bronze armor, the androids react as though attacked. They turn on each other in confused frenzy, hacking at each other with their swords. Then Jason and his companions rush into the fray and finish them off, including some emerging warriors still half-rooted in the plowed furrows.[10]

Recounting this myth more than two thousand years ago, the skeptic Palaephatus (3 *Spartoi*) remarked, "If this story were true, every general would cultivate a field like Jason's!" But the story's dilemma maintains its edge today. How can automaton soldiers distinguish friend from enemy? They could easily turn on each other or on one's allies. How can their orders be recalled or revised? The archaic tale, which some scholars believe predates Homer, is one of the earliest observations that cyborg or robot soldiers will bring problems of command and control.[11]

• • •

The fire-breathing bronze bulls recall the abilities of Talos of Crete, who could heat his brazen body red-hot to roast adversaries (chapter 1). Heated bronze animated statues also bear similarities to some later lore about Alexander the Great. Among the many legends about his military inventions in the *Alexander Romance* traditions, two stand out for deploying fiery bronze statues against enemies. In the first, from the Byzantine-era *Greek Romance*, Alexander devises a strategy to counter the great war elephants of King Porus of India. He heaps onto a large fire all the lifelike bronze statues taken as booty in his conquests. Then his men carefully set out the red-hot statues as their front line on the battlefield. When Porus sends forth his war elephants to attack, the beasts take the bronze men for live soldiers. They crash into the heated metal statues and are badly burned.[12]

The second example presents a more technologically sophisticated version of fire-breathing bulls. In Persian legends that arose about Alexander, the young warlord Sikandar (Iskandar, Alexander) devises an iron cavalry to defeat the army of King Fur of Hind (Porus of India). In some Persian traditions, Alexander is advised by his grand vizier, the sage Arastu (Aristotle, Alexander's tutor). In Firdowsi's epic *Shahnama* (14–15; written in about AD 977, based on earlier oral stories), Alexander's spies make wax scale models of Porus's war elephants to convey how huge and terrifying these unfamiliar beasts are. Alexander then comes up with a battle plan. He commands twelve hundred Greek, Persian, and Egyptian master ironsmiths to forge a thousand life-size hollow iron statues of riders and horses. It takes them a month of painstaking work. The replica horsemen are painted realistically, attached with rivets to saddles, and fitted with armor, shields, and hollow spears. The horsemen's faces would resemble the uncanny, lifelike iron and bronze masks typically worn by Kipchak and other central Asian mounted warriors of the era, which frightened enemies with the impression of an army of metal soldiers. Alexander's craftsmen paint the iron steeds to look like real "dappled, chestnut, black, and gray" horses. The smiths fit the horses with wheels, and then, in the diabolical last touch, they fill the hollow iron figures with volatile naphtha collected from crude petroleum wells.

On the battlefield, Alexander's men ignite the naphtha and set the iron cavalry rolling toward the enemy. The eerie host of metal horses and metal riders, painted to generate the illusion of life, with orange flames shooting from the horses' nostrils and the ends of the riders' spears, create an intimidating juggernaut. Porus's burned elephants run amok; his army is routed. A dramatic color illustration of the spectacle appears in a medieval Mongol version of the *Shahnama*.[13] The statues did not have moving parts but were wheeled like Pasiphae's notorious artificial cow (made by Daedalus, described below).

The iron cavalry evoked a convincing sense of reality mixed with unnatural firepower. The legend reflects practices used by historical Mongol and other nomad armies, who deployed naphtha-wielding cavalry and used the trick of setting dummy soldiers on live horses to make their armies appear larger.[14]

• • •

Since antiquity, human augmentations and enhancements in the form of modern prosthetics have advanced to high levels, from implants, organ transplants, and replacement limbs to neurologically controlled artificial legs and arms. Replacement limbs and bionic body parts—the melding of human and machine—have deep roots in mythology and in actual history. In mythology, for example, the Celtic King Nuada (or Nudd) of the Silver Hand had an arm fashioned by the inventor god Dian Cecht. The Norse goddess Freyja was a kind of "organic cyborg" who combined both flesh and metal. In ancient Hindu epic traditions, the heroine Vishpala lost a leg in battle and Vadhrimati lost a hand—the gods replaced the body parts with, respectively, an iron and a gold replica. In ancient Greek myth, the god Hephaestus made an ivory scapula to replace the hero Pelops's missing shoulder blade.[15]

The earliest historical record of a prosthetic body part was reported by Herodotus (9.37.1–4) in the fifth century BC. Hegesistratus, a Greek from Elis (southern Greece), lost part of his foot under torture by the Spartans. He managed to escape and had a wooden replacement made. He went on to fight in the Battle of Plataea (479 BC) on the Persian side, because of his hatred for the Spartans.[16] Pliny (7.28.104–5) tells how M. Sergius Silus, a Roman veteran of the Second Punic War against Carthage (218–201 BC), recovered from twenty-three wounds and wore an iron hand to replace the one he had lost in battle. The Alexandrian author known as Dionysius Skytobrachion ("Leather-Arm," fl. 150 BC) may have been so named because of a prosthetic arm.

Archaeological discoveries have unearthed surprisingly early evidence of artificial limbs and other body parts, some aesthetic and others functional. A skull from a site in France dated to 3000 BC, for example, sported a prosthetic ear carved from a shell. In Capua, Italy, a skeleton in a tomb of about 300 BC was fitted with a remarkably well-preserved wooden leg covered with thin sheets of bronze. Another skeleton from a grave of the same era, but in Kazakhstan, revealed that a young woman lived several years with a missing foot that had been replaced with the bones and hoof of a ram.[17]

Some of the most sophisticated prosthetic devices are the most ancient. In about 700 BC, a highly skilled artisan who understood human biomechanics made a finely carved artificial toe for a woman whose mummy was discovered in 1997 near Luxor, Egypt. Her replacement toe was not only realistic in appearance; it was tailor-made for her foot and shows evidence of refittings. Worn barefoot or with sandals, her prosthetic toe allowed relatively comfortable mobility: it was constructed in three sections of wood and leather, with a hinge for flexibility.

An ocular prosthesis was discovered by archaeologists in the Burnt City site in Iran. The meticulously realistic artificial eyeball was embedded in the left eye socket of a woman who lived about forty-eight hundred years ago. The anatomical details are amazingly true to nature, with convex surface, cornea, and pupil, and the interior even contained extremely fine golden wires to mimic the capillary network of the eye. The eye was engraved with rays and covered in gold leaf, which would have given the woman an "incredibly striking visage" in life. It is noteworthy that modern attempts to create lifelike prosthetics inspired the robotics engineer Masahiro Mori to suggest the concept of the "Uncanny Valley" in 1970 (for definition and further discussion, see chapter 5 and glossary).[18]

• • •

Some ancient Greek myths tell of those who, like modern military scientists, dreamed of replicating the special powers of animals and birds to amplify human abilities. The artisan par excellence in ancient Greek traditions was Daedalus, the mastermind of facsimiles of life and biotechnological inventions. Since Homer, the word *daedala* denoted any work of marvelous art and workmanship, including those attributed to Daedalus. The chronology and geography of his vast résumé are inconsistent. For example, Pausanias (10.17.4) reported the belief that Daedalus had lived in the mythic "epoch when Oedipus was king of Thebes," while others placed him in King Minos's court about a century before the legendary Trojan War. Various tales locate workshops of Daedalus in Crete, Sicily, and Athens. The activities of the enigmatic, prolific, itinerant "first inventor" called Daedalus can be pieced together from an extensive body of literature and art. The figure of Daedalus takes on a collective persona as a mythic "hero" of invention, the "archetypal craftsman." Was "Daedalus" based on a real person? Modern scholars consider the evolving traditions about Daedalus as attempts to reconcile the many conflicting accounts—and as a reflection of the dual status of Daedalus as both a mythical character and a real historical innovator (or group of inventors) of the remote past.[19]

Unlike Medea's witchcraft that melded *biotechne* with sorcery, Daedalus's cunning devices and human enhancement schemes were achieved with no whiff of magic. Daedalus was a craftsman and inventor, not a magician. Using familiar tools, methods, techniques, and materials, Daedalus deployed creative expertise and technology to achieve wonderful results. Hyperrealistic sculptures, "living statues," were his specialty (chapter 5). But Daedalus is probably most famous for his human-powered flight with wings. And that endeavor started with a witch named Pasiphae. She was Medea's aunt and the wife of King Minos of Crete.

Queen Pasiphae cast a spell on her husband of a particularly foul nature: any time Minos attempted sex with another woman, he ejaculated scorpions, millipedes, and snakes.[20] In turn, Pasiphae was cursed by Zeus with an unnatural desire to have sex with a handsome bull in Minos's herds. She confessed her wish to Daedalus, the brilliant sculptor-craftsman in her husband's court. To fulfill Pasiphae's request, Daedalus constructed a wooden replica of a cow, hollow so that Pasiphae could crawl inside and present herself on all fours for the bull to mount.

This myth was first recounted in writing by the skeptic Palaephatus (fourth century BC) who raised several objections (2 *Pasiphae*). His primary doubt was that a bull would be fooled by an artificial cow decoy, because bulls "smell the genitals of their mates before copulating." But other writers—Apollodorus (*Library* 3.1.4), Hyginus (*Fabulae* 40), and Philostratus (*Imagines* 1.16)—answered that objection, noting that Daedalus covered the wooden facsimile with the hide of a real cow from the herd in the pasture where the bull grazed, so that it appeared and smelled familiar. Modern animatronics experiments have demonstrated that a wide variety of mammals, from meerkats and monkeys to hippos, will interact socially with realistic robotic animals made with actual hides and anointed with species-specific scents. In classical antiquity, there were many anecdotes about paintings and replicas of fauna and flora so accurately rendered that they tricked animals into reacting as though they were alive.[21]

Ancient Greek sources tell of an interesting deception involving a troop of fake "war elephants" that looked and moved persuasively from afar, but failed to convince seasoned warhorses up close. The mastermind was the legendary Assyrian warrior queen Semiramis (probably based on the historical queen Shammuramat, ninth century BC); the story was first recounted by Ctesias (fifth century BC) and then by Diodorus Siculus (2.16–19; first century BC). The numbers are exaggerated but the ruse is plausible. Semiramis, facing a war against a superior Indian army equipped with thousands of war elephants and a strong horse cavalry, ordered her artisans and engineers to slaughter 300,000 black oxen and sew the hides into realistic elephant shapes stuffed with straw. It took two years for the craftsmen, working in a secret place, to manufacture the dummy elephant forms. The ox-hide elephant shapes were then placed over remarkably cooperative camels, and men sat inside to flap the ears and swing the trunks in naturalistic fashion. Semiramis expected to gain the advantage because the Indians believed that only their armies deployed elephants. Indeed, the Indian commander was taken aback to see the "multitude of war elephants" approaching the battlefield. His cavalry, being quite used to elephants, attacked boldly. But upon reaching the fake elephants, the horses shied and ran amok when they detected the unfamiliar odor of the hidden camels.

Several instances of realistic fake animals were reported by Athenaeus. He told of male dogs, pigeons, and geese that attempted to copulate with female replicas of their species. One example was a bronze cow so seductive that it was mounted by a real bull at Priene, a town on the coast of Asia Minor (Athenaeus *Learned Banqueters* 13.605–6).

The sensational myth of Pasiphae mating with a bull is one of several myths about biotechnology allowing humans do things beyond what ordinary humans can (or should) do. Although the replica cow did not have moving parts, it was an imitation of life convincing enough to attract a real bull to mount it when it was wheeled out to the pasture. Daedalus's realistic, life-size sex toy presents a remarkable form of ancient *techne*-pornography. The witch-queen Pasiphae's lust for a bull is nothing like the fanciful liaisons, never explicitly detailed, between a mortal woman and a god in animal disguise, such as Zeus in the form of a swan impregnating Leda. The cow made by Daedalus was not an automaton or machine; rather, in effect, Pasiphae became the internal "living" component of a "sexbot" heifer fabricated with the intention of enabling her to copulate with a live bull. The details in the myth of Pasiphae's zoophilia compel one to visualize the grotesque sex act made possible by Daedalus's cunning biomimetic design.[22]

The story of how Daedalus enabled Pasiphae's bestiality was very popular in Greek and Roman times, perpetuated by many ancient authors.[23] Illustrations of the Pasiphae tale abound in frescoes, mosaics, sarcophagi, and other artworks. A relief on a clay cup made in Tarsus, Anatolia, in the first century BC, for example, depicts Daedalus showing Pasiphae the lifelike heifer. Daedalus presents the cow to Pasiphae in several colorful frescoes discovered in Pompeii and Herculaneum (in one of the paintings, Daedalus's bow-drill is shown). A similar scene appears in the mosaic floor of a Roman aristocrat's villa in Zeugma, Asia Minor. The story struck chords in the Middle Ages and later times too. Medieval miniatures tend to focus on the romance shared by Pasiphae and a gentle, love-struck bull, while modern paintings and etchings often show a lascivious Pasiphae eagerly entering the wooden cow.[24]

As Palaephatus pointed out, what happened next in the myth would have been impossible because different species cannot reproduce offspring, and, moreover, no woman could tolerate sex with a bull or carry a fetus with hooves and horns. In the myth, Pasiphae gives birth to a monster: a baby boy with the head of a bull. The question of how Pasiphae could breastfeed the infant Minotaur arose in antiquity, with some suggesting that a real cow would have to have been his wet nurse. A fine red-figure painting on a cup of the fourth century BC found in an Etruscan tomb shows a frowning Pasiphae with the baby Minotaur on her lap (fig. 4.3). Her hand gestures suggest surprise or hesitation. The earliest artworks depicting the Minotaur antedate the written myth by centuries, going back to the eighth century BC, and by the sixth century BC the Minotaur had become a favorite subject for vase painters.[25]

The Minotaur's birth was a nasty shock for King Minos. Another branch of the myth tells how the Minotaur—who grows up to be a cannibalistic ogre—is imprisoned in the Cretan Labyrinth, a bewildering covered maze designed by Daedalus, of course. Every year a group of young men and maidens from Athens must be sacrificed to the Minotaur, until the Athenian hero Theseus manages to slay the man-bull monster in his maze. Theseus escapes from the Labyrinth with the help of Ariadne, daughter of Minos: Ariadne has given Theseus a ball of wool, telling him

FIG. 4.2. Daedalus, with saw, making a realistic cow for Pasiphae, Roman relief, first to fifth century AD, Palazzo Spada. Photo by Alinari.

FIG. 4.3. Pasiphae and the baby Minotaur, red-figure kylix found at Vulci, fourth century BC, Cabinet des Medailles, Paris. Photo by Carole Raddato, 2015.

to tie one end to the entrance of the Labyrinth and unroll the yarn, so that after killing the Minotaur he can follow the thread, retracing his steps. It is none other than Daedalus who has given Ariadne the ball of wool and the instructions for threading his own Labyrinth.[26]

Deeply offended by the inventor's crimes, Minos imprisons Daedalus and his young son, Icarus, in the Labyrinth. What escape plan would Daedalus devise?

⬢ ⬢ ⬢

Gazing at the horizon where sky met sea, Daedalus dreams up an audacious scheme to free himself and his son from Minos's prison. What if they could fly away like birds? The myth of Daedalus and Icarus soaring aloft on wings made from real feathers and wax is another case of imaginary biomimetic technology to enhance human powers. Narrated by so many storytellers over the centuries, memorialized by countless artists, the tale is one of the most beloved myths of classical antiquity.[27]

Daedalus collects bird feathers and layers them according to size like real pinions, using beeswax (or glue, one of his inventions). He makes two pairs of wings to strap onto himself and his son. Daedalus instructs Icarus to be careful not to fly too high, lest the sun's heat melt the wax or glue, and to avoid dipping too low over the sea, because the moisture might cause the wings to fall apart. But young Icarus, enraptured by the experience of flight, soars too high. As the sun melts the wax, the feathers flutter away and the youth plummets into the sea.[28]

In sorrow, Daedalus flew on, stopping at various Mediterranean islands, and finally making his way to Camicus, Sicily, ruled by King Cocalus. Some said Daedalus dedicated his wings to Apollo in a temple at Cumae, whose walls were decorated with the inventor's life story painted by Daedalus himself. Some skeptical writers, such as Palaephatus (12 *Daedalus*) and Pausanias (9.11.4), rejected the myth of his flight, however. They suggested that the story arose because Daedalus was in reality the first inventor of sails, which archaic people had once likened to wings that allowed ships to "fly" over the waves. In this story, Icarus drowned at sea and was buried by Heracles on the island of Icaria.[29] But the main thread of the myth continues with King Cocalus welcoming Daedalus and offering him protection from Minos. Everyone knows that

FIG. 4.4. Daedalus making wings for Icarus at his workbench, ancient Roman relief, Museo di Villa Albani, Rome, Alinari / Art Resource, NY.

FIG. 4.5. Icarus with wings, small bronze figure, about 430 BC, inv. 1867,0508.746. © The Trustees of the British Museum.

FIG. 4.6. Icarus flying over fisherman in boat; King Minos in the city of Knossos. Roman lamp, first century AD, inv. 1856,1226.470. © The Trustees of the British Museum.

the king of Crete is pursuing his escaped captive, looking for Daedalus in all the major cities across the Mediterranean.

The earliest references to the escape from Crete by human-powered flight are not written but artistic. The oldest image, discovered in 1988, is fascinating for two reasons. It is Etruscan, not Greek, evidence that the Daedalus flight legend had already reached Italy by word of mouth by the seventh century BC, long before the myth was first written down. On an Etruscan *bucchero* jug made in about 630 BC a winged man is labeled

"Taitale," Daedalus's name in Etruscan. On the other side is Medea with her cauldron, inscribed with her Etruscan name "Metaia." This unique pairing of Daedalus and Medea is unparalleled in ancient art; it suggests that the Etruscans linked these two mythical figures because of their wondrous *biotechne*.

Many Etruscan carved gems depict Daedalus/Taitale at work. Another unusual Etruscan artifact, a beautiful golden *bulla* (locket, 475 BC) is decorated with images of Daedalus and Icarus on each side, labeled with their Etruscan names, Taitale and Vikare. They are wearing their wings and carrying tools (saw, adze, axe, and square), details that emphasize craftsmanship and technology.

FIG. 4.7. Daedalus hovering over the body of Icarus fallen on the shore, an eighteenth-century drawing of an ancient mural, Pompeii, first century AD. Ann Ronan Picture Library, London, HIP / Art Resource, NY.

The earliest Greek representation of Daedalus is on a vase of about 570 BC: he is wearing wings and carrying an axe and a bucket. The earliest confirmed image of Icarus is on a fragment of black-figure Athenian pottery of about 560 BC showing the lower half of a man with winged footgear, clearly labeled "Ikaros" (wings on his feet appear in other ancient artworks too). A painted red-figure fragment of about 420 BC shows Daedalus fastening the wings on Icarus, and on a fifth-century BC vase, Icarus plunges into the sea. On a fragment of a fine red-figure vase (390 BC, fig. 4.8) we see a devastated Daedalus carrying his dead son.[30]

FIG. 4.8. Daedalus carrying his dead son, Icarus, Apulian red-figure pottery fragment of a krater, Black Fury Group, about 390 BC, inv. 2007,5004.1. © The Trustees of the British Museum.

More than a hundred ancient images of Icarus and Daedalus are known. Many of them show Daedalus at work surrounded by his tools, making the wings; others show him attaching the wings to his son, Icarus, and Icarus falling from the sky. In Roman times, the story continued as a favorite poignant subject for artists, appearing on carved gems, molded clay lamps, bronze figurines, reliefs, and frescoes. A large group of Roman cameos and glass gems illustrate the story, while several murals in Pompeii capture the moment of Icarus's death, with a horrified Daedalus hovering above his son's broken body on the seashore. The myth's merging of optimism and despair made it a compelling allegorical topos in the Middle Ages too. Although the story may seem a cliché today, one can appreciate how it may have been read: high hopes for man-made technology to artificially enhance human capabilities are cruelly dashed by complacency, hubris, and unanticipated consequences.[31]

• • •

Yet the dream that men could somehow fly like birds far above the earth did not die with Icarus. After all, in the myth Daedalus and Icarus did become airborne and flew successfully, and—despite the high cost of his innovation—Daedalus himself survived the flight to Sicily. Humans hitching rides on birds and insects are featured in Aristophanes's plays, in Aesop's fables, and in ancient Persian traditions. A unique ancient "science fiction" about human flight was written by Lucian of Samosata (b. ca. AD 125). In *Icaromenippus* (or "The Sky Man"), Lucian's popular tale, the philosopher Menippus imitates Daedalus and makes himself a pair of wings to fly to the moon. On his voyage, he observes that earthlings resemble tiny ants scurrying about meaninglessly.[32]

One of the most memorable flying "machine" designs in antiquity appears in the *Alexander Romance* legends, in which Alexander is consumed by the desire to explore two great unknowns, the heavens and the oceans. He harnesses the power of birds to fly and dives like a fish in the deep sea, thanks to two inventions. One device is decidedly magical but the other involves technological ingenuity.

Alexander's diving bell required creative technology. His discovery of a huge crab and giant pearls on a beach fuels Alexander's wish to explore the mysterious depths of the ocean and see its denizens for himself. In classical Greece, primitive diving bells, described by Pseudo-Aristotle (*Problems* 32.960b32), already allowed deep-sea sponge divers to remain under water longer by breathing air trapped in an upside-down cauldron let down into the sea. In the *Romance* legend, Alexander explains how he made a diving bell by encasing a large, man-sized glass jar inside an iron cage, sealed by a lead lid. Alexander climbs inside. Breathing the air trapped in the glass vessel, he is lowered into the ocean by a chain from his companions' ship. At a depth ranging between 454 and 1,400 feet depending on the version, Alexander observes many fabulous deep-sea creatures.

But he almost does not survive the expedition. Suddenly a gigantic fish seizes the diving bell, dragging it and the ship along more than a mile. The great fish crushes the iron bars in its jaws, and finally spits the glass vessel with Alexander still inside onto the beach. Gasping on the shore, Alexander tells himself to give up "attempting the impossible!"[33] As with the fall of Icarus, the "moral" often attached to the *Romance* traditions cautions against the hubris of overreaching human limits. But, in fact, the thrilling audacity of Alexander's undersea and space adventures—to go where no human had gone before—seems more likely to obscure that message. Despite the risks, like Daedalus the bold explorer did live to tell the tale.

Pictures of Alexander "piloting" his diving bell and his flying machine appear in literally hundreds of illustrations in manuscripts, mosaics, sculptures, and tapestries from AD 1000 to 1600. Unlike the technological construction of his iron and glass diving bell, the flying machine is powered by two huge unidentified white birds, vultures, or griffins, goaded ever upward by horse livers dangled on spears above them. The fantasy plays on the folk motif of the donkey lured forward by a carrot on a stick.[34] Alexander flies higher and higher and the air becomes colder and colder, until he peers down at the earth, which now resembles a small globe in the blue ocean's bowl, seemingly insignificant compared to the vastness of the heavens. The scene is remarkably prescient, anticipating the humbled reactions of modern astronauts and viewers of the first pictures of the small blue planet Earth seen from space. This story elaborates on Alexander's wishes to surpass the limits of human capacities, seeking knowledge "beyond the world." Satisfied with his bird's-eye perspective from the stratosphere, Alexander returns to earth.

Daedalus too returned to earth. As we saw, he landed in Sicily and found refuge from King Minos in the court of King Cocalus of Camicus. We pick up the thread of this peripatetic inventor's exploits in the next chapter.

HUMAN-POWERED FLIGHT

The experiments by Daedalus and Alexander reflect an age-old fascination with technology's potentials, envisioned in early myth, legend, and folklore, to exceed human boundaries and create artificial human enhancements. The wish to imitate birds' exhilarating freedom persists, leading many others to try to achieve Daedalus's feat. In the Greek myth, Daedalus's "impossible" human-powered flight involved simply imitating birds, by flapping man-made feathered wings attached to one's back and arms. Large kites in the shape of birds' wings and other wing-beating flying devices were tested in China as early as the first century AD.[35] A Chinese text of the fourth century AD relates that a people of the Far West invented a flying machine driven by wind and had to make an emergency landing in Shang dynasty territory (Yellow River valley, ca. 1600–1046 BC). The Shang ruler destroyed the machines so that they could not be copied, but the stranded pilots rebuilt them and flew back home.[36]

In about 1500, Leonardo da Vinci, who was familiar with Greek myths, not only made designs for a diving bell and suit, but also sketched several plans for human-powered ornithopters (mechanical wing-flapping devices modeled on bird and bat wings). There is no evidence of physical prototypes or test flights for Leonardo's plans. But models based on his drawings have been made, most recently in 2006 by the Victoria and Albert Museum in London for an exhibit on early flight.

The glorious notion of flying by human power alone has inspired numerous intrepid modern inventors to find ways to overcome the problems of aerodynamics and power-to-weight ratio. One clever suggestion was to find a way to use foot-pedaling energy. Leg-powered flight was long considered to be impossible. Aeronautical engineers believed that no aircraft could be light enough to fly on such a limited source of power and yet be robust enough to carry a pilot—who of course would have to possess extraordinary strength and endurance. One of the first attempts was a "cycleplane" built in 1923, but it achieved only twenty-foot hops. In 1977, advances in strong, lightweight materials resulted in a human-powered plane flown by a cyclist-hang-glider pilot, who reached the modest altitude of ten feet and flew just over a mile.

It's diverting to speculate on some potential practical options that existed in antiquity for the mythic Daedalus, such as kites or glider sail-wings. Chinese chronicles record that a prisoner named Yuan Huangtou unwillingly soared about one and a half miles with an owl-shaped kite in about AD 559, a primitive approximation of uncontrolled "hang gliding" (chapter 9).[37] Notably, in some ancient Greek traditions Daedalus was credited with the invention of sails for ships. Coarse linen with high tensile strength was used for sailcloth in Minoan Crete, known for its fine spinners and weavers. Linen sailcloth could be waxed for waterproofing. The natural materials and technical skills to make a simple glider were available in antiquity. A simple, experimental glider design could have been constructed by stretching and gluing waxed sailcloth over a lightweight wicker framework of giant reeds (*Arundo donax*), similar to the working gliders made by aeronautics pioneer Sir George Cayley (1773–1857), who tested his ideas with small models before building larger ones.

In myth, Daedalus was associated with weavers' and spinners' balls of thread. In antiquity the membraned wings of bats captured attention, and spiders were admired for floating on fine silk gossamer and weaving strong silken webs. Venturing for a moment into an ancient realm of science fiction to imagine an alternative myth for Daedalus, one might picture the inventor weaving tensile spiderwebs to make a lightweight sail-wing apparatus, a kind of ancient glider.

Early modern versions of modern hang gliders were hindered by low lift-to-drag ratios, but now, thanks to aluminum alloy and composite frames covered with ultralight laminated polyester films, hang-glider pilots can soar for hours on thermal updrafts at altitudes of thousands of feet, simply by shifting their body weight, with little exertion, imitating the dynamic soaring ability of albatrosses and shearwaters. With a modern hang glider and the help of the winds, a Daedalus could island-hop from Crete to Sicily.

In 1988, inspired to replicate Daedalus's flight pattern in the Aegean, the Greek Olympic cycling champion Kanellos Kanellopoulos skimmed over the Aegean Sea from the island of Crete to the island of Santorini in an ultralight craft, *Daedalus 88*, propelled by pedals. His record-setting flight of 72 miles, at an altitude of 15–30 feet, took about 4 hours of intense pedaling. The experiment was sponsored by the MIT Department of Aeronautics and Astronautics. In 2012, the Icarus Cup was established by the Royal Aeronautical Society in England, to promote the sport of human-powered flight. How amazed Daedalus would be, if only he could witness the continuing legacies of his epic flight to freedom.[38]

CHAPTER 5

DAEDALUS AND THE LIVING STATUES

AFTER HIS SAFE arrival in King Cocalus's court, Daedalus's mythic biography continued as he resumed his role as an architect, artist, and engineer in Sicily. According to ancient local traditions, Daedalus designed an impregnable acropolis for Cocalus at Acragas (founded in about 582 BC, now Agrigento). The summit could be reached only by a narrow, circuitous passageway, an echo of the Labyrinth in Crete. So ingenious was the plan that the fortress could be defended by just three or four men. Temples to Apollo at Cumae and Capua were also ascribed to Daedalus, among numerous other architectural works scattered across the Mediterranean from Egypt to Libya.

Daedalus also spent time in Sardinia during his flight from Crete. The mysterious stone towers, the *nuraghe* of the Nuragic era (tenth to eighth century BC) dotting the island of Sardinia, were thought to be of his design. Sardinia is also the home of the enigmatic Nuragic stone giants of Mont'e Prama (chapter 1, fig. 1.8), which scholars compare to so-called Daedalic-style statues on Crete made in the seventh century BC. Archaeologists point out that advanced tools, surprising for an archaic culture, were used to carve the stone giants of Sardinia. This might help to explain why Daedalus was linked to the island. The statues show evidence of the use of sophisticated metal implements such as stone chisels with different sized blades, hand scrapers, the drypoint stylus, and grooved tooth chisels (which were not introduced in Greece until after the sixth century BC). As mentioned in chapter 1, the striking robot-like faces of the statues follow a "T-scheme" with pronounced brows and nose over eyes rendered with two concentric circles and a slit mouth. Making those perfect concentric circles required technological skill using a compass—and, indeed, archaeologists have discovered Nuragic drills and a complex iron compass on Sardinia.[1]

• • •

For King Cocalus in Sicily, Daedalus devised a cantilevered platform for the Temple of Aphrodite on a precipice at Mount Eryx. To honor the goddess of love, Daedalus was said to have created a gilded ram whose horns, hooves, and woolly body were "so perfect that it would be taken as an actual ram." The celebrated Bronze Ram of Syracuse, one of a pair from the palace of the tyrant Agathocles of Sicily (chapter 9), gives an idea of what the ram ascribed to Daedalus might have looked like (fig. 5.1, plate 6). Another marvel among the rich treasures in the Temple of Aphrodite at Mount Eryx was a perfect honeycomb made of gold.[2] Both objects were of such splendid artistry that they were naturally attributed to Daedalus.

The imitation golden honeycomb was an amazing artifact. How could a human craftsman capture all the details, texture, and geometry of such a fragile, ephemeral natural object in permanent metallic form?

The British artist Michael Ayrton (1921–75) was devoted to re-creating some of the legendary wonders attributed to Daedalus. Working with a goldsmith, Ayrton demonstrated that the fabrication of a delicate golden honeycomb—although laborious and requiring great skill—was "a far less miraculous achievement to a metal worker than to an historian." Historians, he noted, tend to underestimate the ingenuity and technological expertise of ancient artisans.[3]

The lost-wax technique of casting metals, described in chapter 1, could employ as the core a natural object, such as a pinecone or shell, allowing an artist to replicate the object with incredibly precise details. Ancient Egyptian goldsmiths first perfected the painstaking process. We know that Egypt carried out lively trade with Minoan Crete, so Greek craftsmen might well have learned the technique at an early date. As Pliny (33.2.4–5) remarked in his discussions of intricate gold-working skills, "Man has learned to challenge nature!" In *The Maze Maker*, Ayrton's remarkable novel channeling the mythic inventor, he describes the casting process of the honeycomb, as narrated by Daedalus. Being made of beeswax itself, the honeycomb serves as its own wax model in

FIG. 5.1 (PLATE 6). Realistic bronze ram. Was the sculptor of this life-size ram inspired by the story of Daedalus's true-to-life ram dedicated to Aphrodite in the time of King Cocalus? Bronze Ram of Syracuse, Sicily, third century BC, Museo Archeologico, Palermo, Scala / Art Resource, NY.

the complex lost-wax process. First he found a real piece of undamaged honeycomb and carefully uncapped each hexagonal cell and drained the honey. Next the honeycomb was meticulously coated with a fine clay slip. To the side of the clay-covered comb, he attached "a tiny pouring cup and thin 'runners' of wax' as vents." Then the object was placed in a kiln until the waxen honeycomb burned away, leaving its exact

FIG. 5.2. Golden honeycomb cast from real honeycomb.

impression in a mold to be filled with molten gold. A perfect golden replica of a real honeycomb was the result.[4]

The structural strength of honeycombs created by "builder" bees was admired by architects in antiquity. For example, in the sixth and fifth centuries BC, marble blocks of temples on Delos and other Aegean islands were carved to resemble massive honeycombs. It is possible that at some point a metal cast of a honeycomb, like the one in the temple at Mount Eryx, might have played a role in inspiring the sophisticated use of hexagonal "honeycomb" cylinders in the construction of stone buildings. The first written mention of this architectural innovation can be traced to mathematical writings of the second century BC. In about 30 BC, the ancient Roman scholar Varro described the so-called honeycomb conjecture, suggesting that the hexagon shape was the most geometrically efficient for compact volume and strength. More than two millennia later, in 1999, Varro's theory was mathematically proven by Thomas C. Hales.[5]

Daedalus's projects for King Cocalus also included innovative waterworks, a network of rejuvenating steam baths. The legend of Daedalus's thermal "spa" is associated with the volcanic thermal springs at Sciacca, near Selinus in western Sicily. Visitors today can still make out the ancient ruins of bathing grottoes that were cleverly constructed to take advantage of the natural hot sulfur springs issuing from the hillside.[6]

The storied career of Daedalus in Sicily was not without drama. King Minos of Crete, as mentioned earlier, was obsessed with avenging the death of the Minotaur. Traveling across the Mediterranean seeking Daedalus, Minos contrived a puzzle to flush out his quarry. The king carried a large spiral seashell with him, offering a fabulous reward to anyone who could string a thread through its convoluted chambers—an obvious allusion to the trick of escaping the great Labyrinth complex built by Daedalus.

When Minos finally arrived in Sicily, he showed the shell to King Cocalus. In hope of winning the reward, Cocalus secretly took the shell to Daedalus. Daedalus placed a drop of honey at the mouth of the shell and drilled a tiny hole at the top. Then he glued a slender thread to an ant and placed the tiny creature in the hole. The ant wound her way through the spirals and emerged with the thread at the mouth of the shell to get the honey. When Cocalus returned the threaded shell to Minos, the king immediately demanded that Cocalus surrender Daedalus, the only person clever enough to solve the puzzle.[7]

Caught out, Cocalus pretends to agree to turn over Daedalus. But first he invites Minos to enjoy a refreshing dip in his highly esteemed hot vapor baths. His guest is attended by the royal princesses, Cocalus's daughters. Readers who recall what happened to men who bathed in rejuvenating hot baths invented by Medea will sense an ominous pattern. Indeed, while soaking in the grotto, Minos is murdered by Cocalus's daughters and Daedalus. They scald Minos with boiling water from the hot springs at Sciacca, an act reminiscent of the fate of King Pelias at the hands of his own daughters and Medea in chapter 2.

The story of Daedalus's sojourn in Sicily and his murder of Minos was told by numerous ancient authors, including Sophocles in his lost play *The Camicians* and Aristophanes in the lost comedy *Cocalus*.[8] The Athenian audiences were quite charmed by Daedalus. According to Athenian lore, after the death of Minos, Daedalus's long, picaresque life continued into its next chapter—in Athens.

● ● ●

As their city grew in prominence, the ambitious Athenians saw a way to enhance their reputation by appropriating Daedalus as their very own star inventor. Legends arose linking Daedalus to Athens. By the fifth century BC, Daedalus had acquired Athenian roots and was said to have created an array of tools, among them the augur, axe, and plumb line. A stylish folding chair was displayed in Athens as his innovation. Daedalus was also given an extensive family tree in Athens. According to the Athenians, the craftsman accepted his sister's young son as his apprentice. His nephew's name, curiously enough, was Talos of Athens.

The Athenian story about this Talos was worthy of a classical tragedy. Young Talos was reputed to be as gifted as his uncle Daedalus. Talos of Athens thought up several brilliant inventions: the potter's wheel, the drawing compass, and other cunning implements. Naturally, the elder Daedalus grew resentful of the young apprentice's accomplishments. The last straw was Talos's invention of a serrated saw. On a jaunt in the countryside, the youth had come across a snake jaw. Playing around with it, he noticed that the row of small jagged teeth cut easily through a stick. Talos created a new iron tool modeled on the snake's teeth. In the Agora, people gathered around to see Talos showing off how well his new tool sawed wood.

In a fit of envy Daedalus murdered his nephew. After pushing him off the Acropolis, Daedalus was discovered secretly burying the body. Athens grieved the loss of their brilliant young inventor: Talos's grave, on the south slope of the Acropolis, was still honored when Pausanias (1.21.4) visited it in the second century AD. According to their myth, the Athenians put Daedalus on trial for murder, and the Council of the Areopagus found him guilty. Daedalus fled Attica and sailed to Crete—where, so the Athenians claimed, he found work with King Minos. According to the new Athenian chronology, this was when Daedalus began his Cretan adventures (described in chapter 4).[9]

• • •

In antiquity, Daedalus's illustrious reputation revolved around his ability to replicate life with staggering authenticity. His specialty was statuary so true to life that the figures were believed to be capable of movement. As noted, the word *daedala* came to describe "Daedalic" wonders, statues and marvelous images so realistic they seemed beyond the scope of human manufacture, apparently wrought by superhuman skills. The list of statues attributed to Daedalus is very long. Besides the ram mentioned above, examples include a pair of tin and brass statues of himself and Icarus on the Electridae islands in the Adriatic; an Artemis at Monogissa, Caria (Asia Minor); a self-portrait statue in the Temple of Hephaestus in Memphis, Egypt; realistic lions and dolphins for an altar on the coast of Libya; and Heracles statues in Thebes and Corinth.[10]

According to a tale recorded by Apollodorus (*Library* 2.6.3), Heracles himself was fooled by Daedalus's spitting-image portrait of Heracles. One night, Heracles unexpectedly came upon the imposing statue in a portico. So startled was the mighty hero that he instantly grabbed a stone and hurled it at the "intruder."

The Athenian playwrights famously drew on ancient traditions and inserted original revisions in their dramas about mythological events and characters. Daedalus's myth was no exception. Daedalus's so-called living statues were featured in numerous Athenian plays, now known only from fragments quoted by other authors. We know that Sophocles and Aristophanes each wrote a play called *Daedalus*. In both plays, characters claim that Daedalus's animated statues must be bound in place or they will escape. In Euripides's extant play *Hecuba* (ca. 420 BC) Daedalus's automata are compared to those made by the god Hephaestus, and his comedy *Eurystheus* also refers to *daedalic* animated statues. A comic play by Cratinus (*Thracian Women*, ca. 430 BC) jokes that a bronze statue that runs away was made by Daedalus, and a fourth-century BC comedy by Philippus features a wooden statue carved by Daedalus that can speak and walk. The theme of runaway statues became a popular Athenian joke, taken up by Socrates (chapter 7). Artists employed the theme too. A unique scene of artisans making a horse statue so lifelike that it is chained by the leg was engraved on an Etruscan bronze mirror (discussed in chapter 7, fig. 7.7, plate 8). A group of archaic black-figure vase painters (sixth–fifth century BC) illustrated statues of men and animals on buildings coming to life and escaping their architectural frames.[11]

* * *

Modern scholars have often noted that the figure of Daedalus might originally have been an earthbound human double of the inventor god

FIG. 5.3. The sculptor Phidias making a nude statue, by Andrea Pisano, fourteenth century, Museo dell'Opera del Duomo, Florence, Alfredo Dagli Orti / Art Resource, NY.

Hephaestus. Indeed, the Athenians gave Daedalus a genealogy that made him a descendant of Hephaestus, who was revered alongside the goddess Athena in Athens.[12] A district of Athens came to be named for Daedalus, populated by craftsmen who saw him as their patron and claimed to be his descendants. Socrates, whose father was a stonemason, twice refers to Daedalus as his ancestor.

Socrates also mentions Daedalus in some of his metaphors in Plato's philosophical dialogues. In two instances, for example, Socrates likens vacillating arguments to Daedalus's celebrated moving statues (Plato *Alcibiades* 121a; *Euthyphro* 11c–e). In another passage, Plato's Socrates compares people's fleeting opinions unmoored from reason to Daedalus's animated statues. If one's thoughts or opinions are to be of any value, maintains Socrates, then they—like Daedalus's automata—must be tethered to a plinth, or else they will escape, like runaway slaves (*Meno* 97d–98a).[13]

The ancient Greek comparison of automata to slaves remains a concept with a moral significance in modernity. In antiquity, Greek and Roman masters were held responsible for the behavior of their slaves. Today, prescient philosophers of Artificial Intelligence and robotics ethics maintain that it is imperative that AI and robots be considered tools and property—essentially slaves—and that makers must be held responsible for their programming and behavior.[14]

In about 350 BC, Aristotle discussed automata, puppets, and toys set in motion by artisans' practical *techne* (strings, weights, springs, wheels, and other forms of stored, temporary energy) and their similitude to animal locomotion in his natural history treatises (e.g., *Movements of Animals* 701b; *Generation of Animals* 734b). In a curious passage in *Movements of Animals*, Aristotle, referring to semen as the liquid that "animates" an embryo, draws an analogy to the way "sculptors create statues and automata" that contain latent or potential power akin to wound-up clockwork. Aristotle's discussions allude to legendary animated statues like those associated with Daedalus, but it is also possible that Aristotle had in mind real self-moving machines, "mechanical dolls of some kind" made by contemporary inventors (chapter 9). Notably, Aristotle remarks that "an artifact might imitate" a living thing, and he defines an automaton as "a kind of puppet with the ability to move by itself."[15]

In the *Politics* (1.4, discussed more fully in chapter 7), Aristotle clearly speaks of self-moving statues like those made by Hephaestus and Daedalus. In a complicated passage in *On the Soul* (*De Anima* 1.3.406b), Aristotle specifically mentions Daedalus's self-moving sculptures. The statues come up in his discussion of the atomism theory of the fifth-century BC natural philosopher Democritus (b. ca. 460 BC). Democritus's sixty-some treatises have not survived, but from *testimonia* in other works, we know that he based his theory of living beings and their motion on the existence of minuscule, indestructible, invisible "atoms jostling back and forth." In his comments about Democritus's theory—that ceaselessly moving spherical atoms initiate movement—Aristotle refers to the claim made by his contemporary, the Athenian comic playwright Philippus (mentioned above), that the secret of a famous moving statue of Aphrodite was that Daedalus had poured mercury into the hollow figure. Aristotle's point is to compare Democritus's atomism theory to the way balls of mercury naturally move to draw together.[16]

In fact, the shifting weight of mercury flowing to the end of a tilted tube with enough force to change the center of gravity was used to animate self-moving toys in medieval and early modern times. The engineer Heron of Alexandria (first century BC) designed self-opening doors for temples using boiling water and pulleys, and he stated that others used an alternative system based on heated mercury. It is not implausible that mercury could have been used in antiquity to animate devices. The idea that the little-understood metallic fluid called "quicksilver" or "living" mercury could impart mobility to a statue also appears in ancient Indian texts about automatically moving machines. For example, a light wooden model of a giant bird "flew by the energy generated from vats of boiling mercury," and mercury was the key substance to power a sort of perpetual-motion machine.[17]

● ● ●

According to a brief poem by Pindar (*Olympian* 7.50–54, written in 464 BC), a group of legendary animated statues with similarities to works by Daedalus were located in Rhodes. "All along the avenues," wrote Pindar, stood works of exalted art so gloriously crafted that they seem to "breathe and move." An ancient scholiast's commentary on the poem calls the statues "moving things with a soul or life spark." In this case, the maker was not said to have been Daedalus or Hephaestus, but the Telchines, blacksmith wizards of magical metallurgical lore, fabled to be the original inhabitants of Crete and Rhodes. The Telchines carried out activities similar to those of Hephaestus, but on a smaller scale, forging weapons and baubles for the gods. The powers of the statues of Rhodes recall the bronze guardians defending harbors and borders, the function of the mythic Talos of Crete and the historical Colossus of Rhodes (chapter 1).[18]

The legendary "living statues" attributed to Daedalus are of great interest as examples of imaginary and genuine "artificial life" described by classical writers. Many claimed that *daedala*, life-mimicking sculptures, could move their eyes and make sounds, lift their arms, and take steps forward. At the same time, however, controversy arose over the nature of "living statues." Could Daedalus's statues really move on their own? Or was their movement illusory? Numerous ancient Greek accounts refer to wood, metal, and marble statues that could move their heads, eyes, or limbs, perspire, weep, bleed, and make sounds. The archaic idea that statues, especially of divinities, possessed agency has a deep history, long before the fifth and fourth centuries BC when artists began to create exceptionally lifelike figures and historical inventors began to design self-moving devices (chapter 9). It was possible to make statues with parts and hidden or internal mechanisms capable of movement, such as nodding, moving inset eyes, raising arms, opening temple doors, and so on. Hollow statues with cavities and tubes allowed priests to ventriloquize their voices, and Plutarch, Cicero, Dio Cassius, Lucian, and others discuss ways to cause a statue appear to shed tears, sweat, or bleed.[19]

Some writers, such as Diodorus Siculus (4.76), maintained that Daedalus must have "towered above all others in building arts, metal and stone work," and crafted "statues so like their living models that people felt that they were somehow endowed with life." Others proposed that Daedalus was the first sculptor to depict the walking pose in art. "This is the workshop of Daedalus," wrote Philostratus (*Imagines* 1.16); "all around are statues, some with forms blocked out, others in a quite complete state in that they are already stepping forward and give promise of walking about. Before the time of Daedalus, the art of making statues had not yet conceived such a thing."[20] On the other hand, writing in the same era (third century AD), Callistratus (*Ekphrasis* 8) described fourteen well-known bronze and marble sculptures, and he attributed the motion of Daedalus's statues to some sort of "mechanical" workings (*mechanai*).

Whether or not statues made by the mythic inventor Daedalus could actually move is moot. What matters is how they were described and envisioned in antiquity. Some historians and philosophers of science argue that myths about Talos and other literary accounts of "living statues" cannot be taken as evidence that people "imagined the building of mechanical automata" in antiquity—because mechanical conceptions cannot be envisioned before the technology actually exists. Berryman's study of mechanics in ancient Greek thought takes a literalist view of imagination and innovation: "We should not expect people to be able to *imagine* what devices can actually achieve without practical experience." In this admittedly "tautological" view, no one in antiquity could have "*imagined*" such inventions "unless they were informed by experience with technology" to compare them to. In other words, there must have already been some "technology available" before anyone could have conceived of the techniques or tools that might achieve the results described in the myths.[21]

There are of course tensions and gaps between imagination and actuality, representation and reality. Yet it seems obvious that the long history of human innovation relies on the ability to imagine or contemplate unheard-of technologies beyond what already exists or is possible. Indeed, the ancient Greeks are generally acknowledged as innovators in culture, literature, politics, philosophy, the arts, warfare, and science; they embraced creativity, novelty, and imagination. Instead of assuming that changes, improved techniques, and new technologies somehow simply *happen*, ex nihilo, the Greeks saw dreams, ambition, inspiration, resourcefulness, skill, effort, competition, and ingenuity as the essential drivers of change and invention in all fields of endeavor. They could, in literature and art, imagine all manner of things that "could happen." Not all creativity is based on technological precedent or material resources. It is because of surprising ideas and "novelty in the ancient Greek imagination and experience" that "saliently different" concepts and innovations "emerge into being," remarks Armand D'Angour in *The Greeks and the New*. Moreover, imagining technologies that do not yet exist has always been the wellspring of the genre of speculative fiction that we call "science fiction" today, which modern Greek and Latin scholars have traced back to classical antiquity. "Where science fiction leads, philosophers and inventors soon follow."[22]

The animated figures and artificial human enhancements made with prodigious creativity and expertise using familiar materials, tools, and technology to achieve amazing results, as described in classical traditions, are not *literal* prototypes of modern, full-fledged robots and other forms of man-made life. As noted earlier, their internal workings are inscrutable, expressed in mythic language, rendering them "black boxes." But they are significant to us because the accounts show that people in antiquity *could* imagine artificial life and speculate on its possible realization through some ingenious, sublime *biotechne* not yet known or understood. The myths express the idea that there might be discoverable practical ways to achieve synthetic nature in the forms of humans or animals; that perhaps there were ways to create artificial life outside or beyond mere magic or fiat.[23]

A striking aspect of the stories of "living statues" is that ancient philosophers, poets, and playwrights tell us that contemporary images and sculptures of startling realism called up conflicting strong emotions in the viewers.[24] By the fifth century BC, Greek sculptors were achieving extraordinary levels of anatomical verisimilitude, with exceedingly minute details of veins and musculature and a variety of facial expressions. Sculptors began to depict naturalistic, fluid poses that had been impossible before innovations in artistic technology. And keep in mind that both marble and bronze statues were realistically painted. A host of eminent artists' works were described by Pliny.[25] Among his examples of sculptures of "miraculous excellence and absolute truth to life" was a bronze dog licking its wound—a statue so valuable that it could not be insured for loss but had bodyguards charged to defend it with their lives. Pliny also singled out Pythagoras of Rhegium (fifth century BC), who was renowned for his muscle-bound marble athletes with visible tendons and veins. The festering ulcer on the leg of his Lame Man caused viewers to wince with sympathetic pain. The paunchy and balding

FIG. 5.4. Athena visiting the workshop of a sculptor (Epeius?) making a realistic horse statue (Trojan Horse?). Attic red-figure kylix, by the Foundry Painter, about 480 BC, Staatliche Antikensammlungen and Glyptothek Munich, photographer Renate Kühling.

portrait statues made by the Athenian sculptor Demetrius of Alopece (ca. 400–360 BC) were so "lifelike that they were unflattering."[26] People even developed the desire to have sex with erotically compelling statues (see chapter 6).

Meanwhile, painting masterpieces began to feature astonishing depth and perspective. Compelling three-dimensional effects made hands and objects appear to project from the surface. Examples from the fourth century BC, described by Pliny in his *Natural History*, include Aristides of Thebes who painted emotions so skillfully, and Apelles, whose life-size pictures of energetic horses elicited neighs from live horses. Several ancient writers praised the works of Theon of Samos, who specialized in "imaginary visions that they call *phantasias*," vivid paintings accompanied by 3-D and theatrical effects of sounds, music, and lights to give realistic "sense-surround" impressions. Another great Greek artist was Parrhasius, whose incredibly lifelike portraits of athletes appeared to pant and sweat. For his vivid painting of Prometheus ravaged by the Eagle of Zeus, it was whispered that Parrhasius must have tortured a slave to death as his model. The paintings of Zeuxis, Parrhasius's rival, were examples of unprecedented illusionism. These and other artists competed with each other to produce astounding trompe l'oeil paintings and objects, such as luscious-looking bunches of grapes that deceived birds into trying to peck them.[27]

As we will see in chapter 9, by the Hellenistic era a number of artisans were designing and making real mechanical models of humans and animals, such as serving maids, whistling birds, moving serpents, drinking horses, and so on. Marvels of artificial life only imagined in the ancient myths were being realized in engineering plans and inventors' workshops.

As artist Michael Ayrton noted, modern historians tend to undervalue the role of technical ingenuity in ancient artworks. In his survey of realistic artworks, Pliny explained how bronze sculptors made lifelike plaster (and wax) casts of living people, a technique that enhanced the realism of portraits. Physical evidence for the use of plaster and wax casts of real people's bodies to make phenomenal, true-to-life bronze sculptures has come to light in some magnificent statues of the fifth century BC. These unexpected discoveries of artistic technology shocked the modern art world; we had been accustomed to assuming that classical sculptors possessed inimitable, awesome virtuosity in achieving such realism in

FIG. 5.5 (PLATE 7). Realistic bronze and marble statues. Upper left, face of the Hellenistic bronze Boxer of Quirinal (Terme Boxer). Album / Art Resource, NY. Upper right, beard and mouth with silver teeth, Riace bronze statue A, found in bay of Riace, Calabria, Italy, in 1972, thought to be the work of Myron of Athens, 460–450 BC. Museo Archeologico Nazionale, Reggio Calabria, Erich Lessing / Art Resource, NY. Lower left, marble arm of the Discus Thrower, Roman copy of the classical Greek bronze original by Myron of Athens, 460–450 BC. Museo Nationale Romano, Rome, © Vanni Archive / Art Resource, NY. Lower right, athlete, fourth to second century BC, recovered off the coast of Croatia in 1996, Museum of Apoxyomenos, Mali Losinj, Croatia. Photo by Marie-Lan Nguyen, 2013.

their bronze figures. The technique, detected and explained by Nigel Konstam in 2004, helps explain the stunning mimetic qualities of many bronze statues.[28]

• • •

Mercury, quicksilver, was a substance of mystery in antiquity, as we have seen. Curiosity about mysterious lodestones—magnetite, a naturally occurring magnet that attracts iron—led some ancients to suggest that magnets also possessed a kind of life, a soul or breath or daimon within. The strange, rare mineral—popularly called *ferrum vivum*, "live iron"—had bewitching powers to move and enliven objects made of iron. This led creative thinkers to imagine how the stone's inexplicable ability to draw or repel iron might be exploited to mystify viewers. What if "living iron" could allow a human replica made of iron to float in midair, to actually levitate and hover effortlessly like the gods, or soaring birds?[29]

Ptolemy II Philadelphus, the Macedonian Greek king of Egypt (283–246 BC) oversaw many unprecedented engineering feats in Alexandria, including an impressive female automaton (chapter 9). He married his own sister, Queen Arsinoe II, and honored her as a goddess after her death. In 270 BC he decreed that her likeness should grace every temple in Egypt. Pliny reports that the king commissioned a renowned architect to create an especially sublime statue of Arsinoe for a temple in Alexandria. Pliny gives his name as "Timochares," but he may have meant Dinocrates of Rhodes, the brilliant engineer for Alexander the Great, who designed the city of Alexandria and other wonders. The plans called for constructing a vaulted roof of *magnete lapide*, magnetic stone, over a lifelike statue of Arsinoe, either made of iron or with an iron core. The idea was that the queen would miraculously hover unsupported in midair, symbolizing her ascent to the heavens (Pliny 34.42.147–48). Surviving sculptures of Arsinoe are realistic, sensuous portraits, nude or transparently draped, so one can guess a similarly erotic statue was planned for this temple. But the grand project was never completed, owing to the deaths of the architect and Ptolemy II Philadelphus in 246 BC.

In fact, the design for the perpetually or even momentarily hovering Arsinoe was an impossible dream. In his study of the long history of "magnetism fantasies" from antiquity to the Middle Ages, Dunstan Lowe shows how the pervasive lore about "floating statues" arose from misunderstandings of the physics of magnets. "In reality," Lowe points out, Earnshaw's theorem of 1839 remains uncontested to this day: it states that "stable levitation" of a fixed magnetic object "against gravity using only ferromagnetic materials cannot work on any scale." The ancient fascination with magnetic power in third-century BC Ptolemaic Egypt is an example of an attempt to imagine and realize an advanced technology millennia before electromagnetic levitation was perfected .[30]

Yet the vision—the science fiction—of animated statues activated by "live iron" was perpetuated as a kind of "sacred physics" in the ancient world. Over the centuries, numerous reports accumulated, alleging that scores of statues, including likenesses of the Greek-Egyptian god Serapis, the Greek sun god Helios, the mythic Athenian king Cecrops, even a winged Eros/Cupid, really floated in midair, magically suspended or balanced by lodestones. Notably, in the twelfth century AD, a twirling statue of Muhammad, made of gold and silver and presumably iron, was said to have been balanced above a tent by means of four magnets and caused to rotate by fans—an idea that included the concept of rotation, but also impossible. All of these "floating" idols, if they really existed, were supported by other, cleverly hidden means, but they were taken as techno-miracles by viewers and ascribed to ingenious harnessing of magnetism by the learned.[31]

Magnetism as a metaphor for sexual attraction turns out to be an ancient concept. The irresistible, mystical coupling of otherwise lifeless stones, magnetite and iron, was observed in antiquity. The phenomenon was "brought to life" in a pair of erotic statues in a racy Latin poem by Claudian (b. ca. AD 370). The mineral *magnete*, magnetite, writes Claudian, is "animated and invigorated by the hardness of iron" and it "languishes without it." Iron, for its part, is charmed by lodestone's "warm embrace." The poem describes two statues in a temple, a Venus carved of magnetite and an iron Mars, standing some distance apart. The goddess of love and the god of war were lustful lovers in Greek myth: Claudian tells how the priests celebrate their divine love with bouquets and songs. As the figures are slowly moved closer together—suddenly Venus and Mars fly into each other's arms, and it takes effort to pull them apart.[32]

Did these magnetically animated statues really exist in Alexandria, or were they figments of the poet's imagination? Claudian was a native of Alexandria, the home of many magnetic fantasies. The action described in the poem is not impossible levitation but realistic magnetic attraction. One can easily imagine that a pair of small figurines, along the lines of modern magnetic toys, could have been created for entertainment in that sophisticated city of technology.

• • •

Unprecedented innovations and brilliant techniques in Greek art and in mechanical technology, evoked *sebas*, *thauma*, and *thambos*—awe, wonder, and astonishment—in antiquity. Many writers described how people confronted with true-to-life artificial animals and especially facsimile human beings experienced the "shock of the new," a sense of surprise and pleasure—but mixed with acute feelings of disorientation, alarm, and terror. These unnerving effects of artistic illusions, vivid imitations of life, animated sculptures of humans and animals, and statues that seem to actually *be* what they portray can be seen as ancient parallels of the Uncanny Valley phenomenon. The Uncanny Valley, a psychological reaction first identified in robotics in 1970, refers to the unease and apprehension that people experience when they encounter eerie, "not quite but very nearly human" replicas or automata. Anxiety rises steeply when the line dividing the inanimate from the animate collapses, especially with anthropomorphic entities, and actual movement or the illusion of movement intensifies negative emotions.[33]

A genre of ancient and early medieval oral traditions preserved in Hindu and Buddhist literature describes the wonder mixed with fear evoked by superrealistic android robots (*yantra/yanta* "machine, mechanical device" in Sanskrit and Pali, respectively) made by clever machine-makers (*yantrakaras/yantakaras*). The original dates of the oral tales (versions exist in Sanskrit, Pali, Tibetan, Tocharian, Mongolian, and Chinese) are unknown, but the stories began to be committed to writing in the third to first century BC. One tale tells of a brilliant inventor who visits a foreign king accompanied by a lifelike robot that he introduces to the court as his son. The robot, dressed in elegant robes, has "charming manners and dances most beautifully." One day, however, the robot casts flirtatious glances at the queen. The outraged king orders his men to behead the "lascivious young man." The inventor quickly offers to discipline his "son" himself and removes part of the robot's shell to reveal the mechanism inside. Astonished and delighted, the king richly rewards the inventor (see chapter 6 for an ancient Chinese version of this tradition).[34]

The earliest Greek examples of an Uncanny Valley–type response to artificial life occur in Homer's *Odyssey* (11.609–14). In the Underworld, Odysseus reacts with fear when he encounters hyperrealistic images of wild animal predators and murderers with glaring eyes. Odysseus prays that this fiendish artist will not create any more of these terrifying pictures. Later (19.226–30), Odysseus describes an intricately wrought golden brooch depicting a hunting hound mauling a fawn. Everyone marvels at the "living" vignette of the dog seemingly captured in the very act of seizing and killing the fawn as it gasps out its last breath.[35]

In two dramatic instances in lost plays of the fifth century BC by Euripides and Aeschylus, old men are frightened out of their wits by Daedalus's animated statues. In Aeschylus's *Theoroi*, some satyrs are alarmed by effigies of their own heads nailed to a temple. One satyr cries out that they are so real they lack only voices to come alive. Another satyr exclaims that the replica of her son's head would send his mother running and shrieking in horror. Such theatrical anecdotes suggest that classical audiences were familiar with artworks of disquieting realism, and, furthermore, they could imagine an extraordinary artisan who might be capable of even more preternatural mimesis than they had personally experienced.[36]

• • •

Daedalus was imagined in antiquity as a brilliant craftsman, a sculptor of artificial life, and innovator of countless clever tools and designs to augment human abilities. In myth, the inventor not only borrowed the pinions of birds in order to fly to freedom; he was believed to have created such lifelike statues that they moved on their own or at least gave the startling appearance of motion. As mentioned earlier, Daedalus and his works sometimes overlap with those of his divine counterparts, Prometheus and Hephaestus. As we'll see in the next two chapters, many of the marvels wrought by these two divinities eclipse those of Daedalus. Their artifices are still more splendidly "alive" and some even possess "intelligence." Yet both Prometheus and Hephaestus were envisioned using the very same tools, methods, and technologies that the mortal Daedalus wielded in his earthly workshop.

CHAPTER 6

PYGMALION'S LIVING DOLL AND PROMETHEUS'S FIRST HUMANS

THE LIFE AND times of Prometheus, the maverick Titan who deceived Zeus and championed early humans, trace a meandering path in ancient Greek mythology. He is first introduced in Hesiod's poems written in 750–650 BC. Prometheus, enduring his shifting relationship with Zeus, also stars in the fifth-century BC dramatic trilogy *Prometheus Bound*, *Prometheus Unbound*, and *Prometheus the Fire-Bringer*, often attributed to Aeschylus.[1]

Retellings and embellishments of the ancient traditions about Prometheus are found in about two dozen ancient Greek and Latin sources. In the earliest versions, Prometheus was the benefactor of humankind, showing them how to use fire. In later myths his gifts expanded to include speech, writing, mathematics, medicine, agriculture, domestication of animals, mining, technology, science—in other words all the arts of civilization. Of interest in this chapter is the persistent thread of myth describing Prometheus as the creator of the human race, either at the beginning of humanity or after the great disaster known as Deucalion's Flood. This tradition would help explain his concern for humans and his theft of fire for them. The earliest surviving mention of this myth comes from a fragment of Sappho. In about 600 BC, she wrote, "After he created men Prometheus is said to have stolen fire."[2]

The myth of Prometheus making the first people on earth is one of many ancient traditions demonstrating that "human beings were once viewed as artificial creations." Earth and water, combined and brought to life by divine power: this was the earliest human metaphor for life. As in other tales around the world, from *Gilgamesh* to Genesis, the creator or demiurge uses mundane materials—such as clay, mud, dust, bone, or blood—to form male and female shapes that receive the spark of life from gods, wind, fire, or some other force of nature. This mud metaphor would be eclipsed many centuries later, with new understandings of the human body as a mechanistic entity driven by dynamic, moving fluids, and with the invention of mechanical, hydraulic, and pneumatic engineering in the Hellenistic era.[3]

In the ancient Greek myth about Prometheus, the Titan mixes earth and water—or tears—and shapes the mud or clay into the first men and women. By some accounts, he makes all the animals too. Athena is involved in some versions, and in others Zeus commands the wind to give the clay figures the breath of life; other interpretations suggest that fire brought Prometheus's creations to life.[4]

Ancient folklore about Prometheus's creation of the first humans was still circulating when the inquisitive traveler Pausanias toured Greece in the second century AD. He had heard the folklore that Prometheus had accomplished his handiwork near the very old town of Panopeus in Phokis, near Chaeronea, central Greece. Pausanias (10.4.4) visited the fabled site near the ancient town's ruins and saw two large clay boulders in a ravine, each big enough to fill a cart. "They say that these are remains of the clay out of which the whole race of man was fashioned by Prometheus." The "scent of human skin still clings to the large lumps of clay," declared Pausanias. One can only imagine the odor that Pausanias and others detected, but rocks and clays can release distinctive odors when heated, breathed upon, or scraped, owing to chemical composition and trapped gas bubbles.[5]

• • •

A number of Greek tales, as in other cultures' myths, describe lifeless matter, statues, idols, ships, and stones brought alive by gods or magic. These stories of artificial life differ from the tales about the animated statues we have considered so far, such as the bronze robot Talos manufactured by Hephaestus with internal workings and the animated statues attributed to the inventor Daedalus (chapters 1 and 5). In what we might term "magic-wand" scenarios, life is bestowed on inert objects simply by a god's command. No craft or manufacturing processes, internal structure, or notions of mechanics are implied. One example of bringing inanimate objects to life by fiat occurs in the myth of the great flood sent by Zeus. Deucalion and his wife, Pyrrha, are the sole survivors. They learn from an oracle how to repopulate the earth. They each toss stones over their heads, and the stones are immediately transformed into men and women.

The most familiar classical example of a statue magically enlivened by divine order is the myth of Pygmalion and his love for a nude ivory statue of his own making. Ovid's version (*Metamorphoses* 10.243–97) is the most vividly detailed account of Pygmalion. The young sculptor is disgusted by vulgar real women, so he sculpts a virginal maiden for himself. In the modern imagination, his statue is often pictured as marble, but in the myth it is ivory, a warmer, organic medium. His ivory maiden looks so real that Pygmalion immediately "burns with passion for her," caressing her perfect body with awe and desire, imagining that were he to press against her forcefully she would actually bruise. He showers the statue with gifts and words of love. In the Temple of Aphrodite he beseeches the goddess to make his "simulacrum of a girl" come alive.

Pygmalion returns home and makes love again to his fantasy woman's ivory form. To his astonishment, the statue warms to his kiss, and in his embrace her body becomes flesh. Unlike cold marble, ivory is a once-living material with a soft, creamy luster. In antiquity, ivory figures were tinted with subtle, naturalistic colors to resemble real skin tones. Ancient audiences would have imagined her as an exquisitely sensuous, flawless female form. Under her maker's caresses, Pygmalion's statue awakens into consciousness and she "blushes with modesty." Aphrodite has answered his prayer.[6]

It is important to emphasize that Pygmalion's artifact was not constructed to be an automaton. Its realism became reality supernaturally, thanks to the goddess of love. This oft-told ancient "romance" of artificial life takes on new relevance today because it presages ethical questions posed by modern critics of lifelike robotic dolls and AI entities specifically designed for physical sex with humans. "Is it possible," one writer asks, "to have consensual sex with a robot, even one that's aware of its own sexuality?"[7]

Although the Pygmalion myth is often presented in modern times as a romantic love story, the tale is an unsettling description of one of the first female android sex partners in Western history. It is not clear that Pygmalion's passive, nameless living doll possesses consciousness, a voice, or agency, despite her "blushes." Has Aphrodite transformed the perfect female statue into a real live woman, with her own independent mind—or is she now "just a better simulation?" The statue is described as an idealized woman, more perfect than any real female. So Pygmalion's replica "surpasses human limits," much like the sex replicants in the *Blade Runner* films that are advertised as "more human than human."[8] Ovid, notably, does not describe her skin and body as feeling lifelike. Instead Ovid compares her flesh to wax that becomes warm, soft, and malleable the more it is handled—in his words, her body "becomes useful by being used."

Ovid ends his fairy tale with the marriage of Pygmalion and his nameless living statue. He even adds that they were blessed with a daughter named Paphos, a magical feat of reproduction intended to show that the ideal statue became a real, biological woman. Notably, the plot of the film *Blade Runner 2049* turns on a similar magical reproduction of a replicant, the biological birth of a baby to the replicant Rachael, which is supposed to be impossible for artificial life forms.[9]

In retelling the Pygmalion story, Ovid was drawing on earlier narratives, now lost. One source was Philostephanus of Alexandria, who recounted a full version of the myth in his history of Cyprus, written in 222–206 BC. In a variant by the later Christian writer Arnobius, Pygmalion sculpts and makes love to a statue of the goddess Aphrodite herself. No artistic representations of the Pygmalion myth survive from antiquity. But many medieval illustrations show Pygmalion interacting with his ivory statue; the tale served as a kind of prurient religious warning against worshipping idols. By the eighteenth century, European storytellers had finally given Pygmalion's statue a name, Galatea ("Milk-White"). Variations on the Pygmalion myth have proliferated over millennia, inspiring myriad fairy tales, plays, stories, and other artworks.[10]

● ● ●

In the Pygmalion myth, the sculptor's ivory statue is "clearly an artifactual being created for sex."[11] But Pygmalion's ivory woman was not the only statue that aroused an erotic response in viewers in antiquity. There is a long ancient history of *agalmatophilia*, statue lust.[12] Lucian (*Amores*

13–16) and Pliny the Elder (36.4.21) told of men who were passionate for the beautiful, undraped statue of Aphrodite at Knidos. It was created by the brilliant sculptor Praxiteles in about 350 BC, the first life-size female nude statue in Greek art. The men surreptitiously visited her shrine at night, and stains discovered on Aphrodite's marble thighs betrayed their lust. The sage Apollonius of Tyana tried to reason with a man who fell in love with the Aphrodite statue by recounting myths of unhappy trysts between gods and mortals (Philostratus *Life of Apollonius* 6.40). In the second century AD, the Sophist Onomarchos of Andros composed a fictional letter by "The Man Who Fell in Love with a Statue," in which the thwarted lover "curses the beloved image by wishing upon it old age."[13]

In yet another infamous case, reported by Athenaeus (second century AD), one Cleisophus of Selymbria locked himself in a temple on the island of Samos and tried to have intercourse with a voluptuous marble statue, reputedly carved by Ctesicles. Discouraged by the frigidity and resistance of the stone, Cleisophus "had sex with a small piece of meat instead" à la Portnoy.

Most "statue lust" stories feature men having sex with female statues, but several ancient sources relate the sad tale of the widow Laodamia (also known as Polydora) whose beloved husband, Protesilaus, died in the legendary Trojan War. The earliest known text was a fifth-century BC tragedy by Euripides, but the play no longer exists. Ovid's version takes the form of a letter from Laodamia to Protesilaus. They were newlyweds when he departed for Troy (the war lasts a decade). Laodamia aches for her husband's return. Each night Laodamia erotically embraces a life-size waxen image of her husband, who was "made for love, not war." The replica is so realistic that it lacks only speech to "be Protesilaus." Hyginus recounts a variation of the tale. When Protesilaus is killed, the gods take pity on the young couple and allow Protesilaus to spend three precious hours with his wife before he must return to the Underworld forever. Distraught with grief, Laodamia then devotes herself to a likeness—this time in painted bronze—of her husband, showering the statue with gifts and kisses. One night, a servant glimpses the young widow in passionate embrace with the male figure, so lifelike that the servant assumes it is her lover. The servant tells her father, who bursts into the room and sees the bronze statue of the dead husband. Hoping to end her torment, the father burns the statue on a pyre, but Laodamia throws herself on the pyre and dies.[14]

One can compile about a dozen accounts of heterosexual and homosexual love for statues in Greek and Latin sources. Historian of medieval robots E. R. Truitt calls these tales and the story of Pygmalion "parables about the power of mimetic creation" and the ways one can "confuse the artificial with the natural."[15]

Alex Scobie, a classicist, and the clinical psychologist A.J.W. Taylor have pointed out that this particular sexual "deviance" arose at a time when Greek and Roman sculptural artistry was achieving a high degree of realism and idealized beauty. Beginning with Praxiteles, there was "an abundance of sculptured human figures with which people could identify," life-size and very naturalistic in appearance, coloring, and poses. Beautiful, realistically painted statues were not only plentiful but "conveniently accessible" in temples and public places, encouraging "the populace to form personal relationships with them." Nude cult statues were often treated as though they were alive, given baths, clothing, gifts, and jewelry. Writing in 1975, Scobie and Taylor concluded that *agalmatophilia* for marble (or ivory or wax) statues that replicated life with intimate realism was a pathology made possible by the technical expertise of superbly talented artists in classical antiquity. As they and art historian George Hersey, writing in 2009, speculated, advances in anatomically realistic silicone sex dolls and biomimetic, AI-endowed cyber-sexbot technologies will result in the ancient paraphilia evolving into a modern form of "robotophilia."[16]

• • •

Greeks and Romans were not the only ancient cultures to spin tales about sexualized automata. An irresistible female robot appears in a Buddhist tale from the *Mahāvastu* (a collection of oral traditions that were compiled over the period from the second century BC to the fourth century AD). Sanskrit, Tibetan, Chinese, and Tocharian versions of the tradition tell how a celebrated inventor of mechanical devices constructs a lovely, lifelike girl (*yantraputraka*, "mechanical doll") to show off his mastery.[17] The inventor welcomes a foreign guest, a highly respected painter of lifelike images, to his home, and entertains the artist with all manner of honors. That night, the painter retires to his room and is surprised to find a beautiful girl ready to "do service to him." Modest and shy, the girl looks down and does not speak but reaches her arms out to the painter and draws him to her bosom. He notices that a jeweled brooch on her chest rises and falls as though with breath. The painter believes she is a real woman—but who is she? Could she be his host's relative, his wife, sister, or daughter? Or a serving maid? There follows a long passage as the painter weighs the moral risks of having sex with the willing young woman in his room.

Finally the painter gives in to his aroused feelings and takes the girl in his arms with "violent passion." Thereupon the mechanical girl breaks apart, "her clothes, limbs, strings, and pegs falling to pieces." The painter realizes he's been tricked by a cunning artifice. Mortified, he conceives of a way to get even with his host. Taking out his supplies, the artist spends the rest of the night painting a gruesome trompe l'oeil image of himself hanging dead, suspended from a rope on a hook on the wall.

In the morning, the host, fooled by the painted illusion, summons the king and his ministers and citizenry to see the tragic scene of the broken mechanical woman and the painter's suicide. He calls for an axe to cut down the body of his guest. The ruse is revealed when the painter suddenly steps out from hiding and everyone has a good laugh.

The Buddhist story reflects the lifelike realism that was achieved by painters and makers of mechanical androids in ancient Asia (see chapters 5 and 9 for other ancient Buddhist tales about robots). The theme of intense rivalry between the two master artists who trick each other with their creations of preternatural realism is similar to anecdotes related by Pliny (35.36.64–66) about trompe l'oeil contests between the classical Greek artists Zeuxis and Parrhasius (chapter 5). But the Buddhist tale is also a philosophical parable about illusions of self-control and the timeless questions of human free will raised by creations of artificial life. In her study of mechanical beings in ancient Indian literature, Signe Cohen points out that the soulless female automaton stands for the soullessness of all beings, embodying the Buddhist teaching that, in essence, "*We are all robots.*"[18]

• • •

Pygmalion's statue of Galatea is an example of an inert object instilled with life by transcendent love or a god's "supernatural power . . . with no reference to mechanical craft." Accordingly, Minsoo Kang places it in his first category of ancient nonrobots, along with the "biblical story of the creation of Adam and Eve," which was not conceived of as "technological." Indeed, "magic-wand" myths, like the story of Pygmalion, do not involve "mechanical ingenuity" or a "life-imitating machine." But such technological features do distinguish Talos (chapter 1), and they figure in some interesting artistic illustrations of Prometheus as the maker of the first humans.[19]

The tale of Pygmalion's ivory sex doll and the myth about the rolling stones that magically became people after Deucalion's Flood are helpful in distinguishing between unambiguous "magic-wand" tales, like those in Kang's first category, and more complex tales of artificial life and automata that were imagined in mythical accounts that include manufacture using tools and methods, some manner of internal structure, and sometimes even intelligence and agency. In the most familiar versions of Prometheus as an artisan who molds familiar plastic material—clay—into lifelike figures of men and women, a god or goddess bestows the finishing touch that completes the Titan's work. This vision is depicted in widely known artistic illustrations of Prometheus making the first humans, guided by Athena/Minerva who provides the supernatural life spark, symbolized by a butterfly. It is important to note, however, that all of these well-known images were late Roman artworks, created in the early Christian era.

In the late Roman-Christian period, Prometheus as the creator of humans appears in elaborate reliefs on sarcophagi, mosaics, and wall paintings in the third and fourth centuries AD. The images emphasize the collaboration of Prometheus and Athena (Minerva). Prometheus forms small, realistic mannequins of men and women, who lie or stand about awaiting the divine touch to spring to life, much like Pygmalion's statue of Galatea. These scenes have obvious features in common with—and are thought to have influenced—later Christian representations of the biblical creation of Adam and Eve. The popularity of the Prometheus scene on so many Roman sarcophagi may also have represented Neoplatonic concepts of creation in contrast to Christian scriptures about Adam, a religious debate that was ongoing when these scenes were being made.[20]

FIG. 6.1. Prometheus making the first humans, guided by Minerva/Athena, late Roman marble relief, third century AD. Albani Collection MA445, Louvre, photo by Hervé Lewandowski, RMN-Grand Palais / Art Resource, NY.

FIG. 6.2. Prometheus making the first humans, guided by Minerva/Athena. Late Roman marble sarcophagus, third century AD, Capitoline Museum, Rome. Erich Lessing / Art Resource, NY.

Remarkably, however, about a thousand years *before* the Roman-Christian images of Prometheus became so popular on coffins, another group of creative artists in Italy took a very different approach to the fabrication of the first human beings by Prometheus. These Hellenistic-era Etruscan artists illustrated the scene in a way that clearly differentiates the statues magically given life from the creations of Prometheus.[21] On a fascinating group of carved scarabs and seals, the first humans were not imagined as clay dolls awaiting a life spark. Instead the humans are pictured being crafted with tools and assembled piece by piece on a framework, much as a sculptor would construct a human statue beginning with an internal armature or part by part (see fig. 1.9, plate 3). In other words, the gems refer to *biotechne* rather than simple magic deployed to create life.

• • •

Beginning in the fifth century BC intricately carved Etruscan and Etruscan-style gems depicted sculptors and artisans at work, and they illustrated both mythic and real craftsmanship in imaginative ways. Of special interest here are several related miniature scenes, dated to the fourth or third through the second century BC, identified as "powerfully original" depictions of Prometheus creating the first humans. The scenes are engraved on personal rings, seals, talismans, ornaments, and scarabs. Some bear inscriptions (designating the owners) in Latin, Greek, or Etruscan letters. These gems have attracted scant attention despite their extraordinary imagery. The most recent work was by Italian scholar Gabriella Tassinari in 1992; her monograph catalogues sixty-three examples of gems showing Prometheus as the creator, noting differences in style and difficulties of dating. The gems can be divided into two types of scene: in both, Prometheus is shown as a solitary artisan using tools to fabricate the first man (sometimes woman) in a complex, step-by-step process.[22] In the first group, Prometheus forms a human figure in sections on a framework of poles, starting with the head and torso. In the second group, even more surprising, Prometheus begins by making the figure's internal armature—a human skeleton.

How ancient is the idea of Prometheus as the maker of the first humans? Explicit literary references appear in fourth-century BC Greek poems and plays, but the oral tradition appears to be even older.[23] As we have seen, Etruscan artists often interpreted Greek mythological stories in a unique manner on gems, mirrors, and vases (chapters 1–4). The unusual Etruscan scenes of Prometheus (*Prumathe* in Etruscan) might have been inspired by other local oral traditions and art. As Etruscan scholar Larissa Bonfante remarks, "something about Prometheus evidently struck a special chord for the Etruscan artists and their patrons."[24]

In the first type of these engraved vignettes, Prometheus assembles the prototypical human body in sections. Instead of molding clay into human-shaped dolls under the guidance of Minerva, as in the reliefs of the late Roman-Christian era (see figs. 6.1 and 6.2), Prometheus is shown alone, fashioning an unfinished body—usually only the head and torso are complete—supported on a framework of metal or wooden poles. Notably, Prometheus is employing tools and technologies of real craftsmen in antiquity. He uses a hammer or mallet, scraper, scalpel, and "a rod or a rope to measure the proportions of the human figure," and he gauges his work with a plumb line. In figure 6.3, for example, Prometheus uses a plumb bob (plummet and a plumb line) on the incomplete human model attached to poles.[25] In figure 6.4, Prometheus secures a half-formed body to a pole with rope.

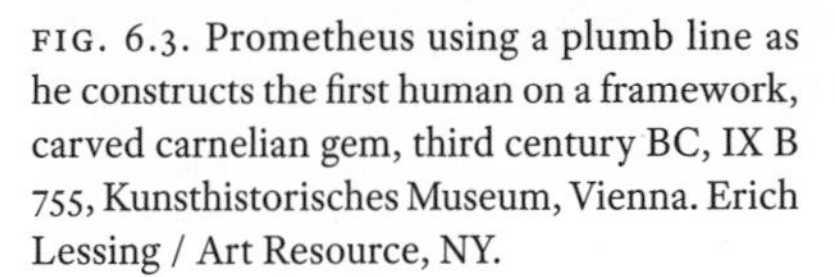

FIG. 6.3. Prometheus using a plumb line as he constructs the first human on a framework, carved carnelian gem, third century BC, IX B 755, Kunsthistorisches Museum, Vienna. Erich Lessing / Art Resource, NY.

FIG. 6.4. Prometheus molding the head and torso of the first man on a frame, sardonyx gem, third century BC. Kunsthistorisches Museum, Vienna. Erich Lessing / Art Resource, NY.

A substantial number of Etruscan and Greco-Roman gems in museum collections have variations of the images in figures 6.3 and 6.4. Some have asked whether the scenes might depict *maschalismos*, the ritual dismemberment of enemy warriors practiced by Etruscans. But when that practice is depicted on gems, we see one or two soldiers using swords to decapitate and sever limbs of foes. Those rare scenes differ dramatically from the set of gems considered here, which clearly show an artisan, typically seated, working with tools to form an incomplete human figure.[26] The pictures of Prometheus building a man in sections recall classical vase paintings of artisans forging and assembling statues of men and horses (see fig. 1.9, plate 3; fig. 5.4; fig. 7.7, plate 8; 7.8, plate 9).

The second type of gems considered here present another striking vision of the process of constructing the first man. In these highly unusual engravings, Prometheus builds the first human being from *the inside out*. He begins his creation with the natural anatomical structure, the skeleton. Skeletons were extremely rare in classical Greek and Etruscan art. As Tassarini points out, however, the main focus of these particular gems is not the skeleton itself but "the creative activity of Prometheus" as a craftsman.[27]

Two gems, dated to the second century BC, once in the collection of Giovanni Carafa, Duke of Noia, are arresting for their depictions of both types of intaglio images of Prometheus making the first man. The gem in figure 6.5 shows Prometheus "working on the modelling of the upper part of a bearded man, supported by two poles." On either side of the scene are the foreparts of a horse and a ram. Their presence reflects ancient versions of the tradition that Prometheus also created the first animals.[28]

The second gem in the Carafa collection, known only by an engraving of 1778, has a curious scene that depicts a partially molded man's torso on a human skeleton instead of on a metal or wood frame. In figure 6.6, Prometheus is seated and holding a tool in his right hand. He is working on the partially molded man's upper back and arms, which are attached to a bare skull and the lower vertebrae, pelvis, and leg bones of the skeleton. The area where the partially fleshed out ribs meet the skeletal vertebrae is similar to the narrow "unfinished" waist in the other gems depicting the upper half of a man. The unfinished man holds a *phiale*, a shallow dish for libations, in each hand.

FIG. 6.5. Prometheus making the first man, flanked by the first horse and ram, second to first century BC. Gem and cast © Collection of the Duke of Northumberland and Beazley Archive, Oxford University; photo by C. Wagner. C. Engraving, *Alcuni monumenti del Museo Carrafa* (Naples, 1778), plate 25. Courtesy of Getty Research Institute, Los Angeles (89-B17579).

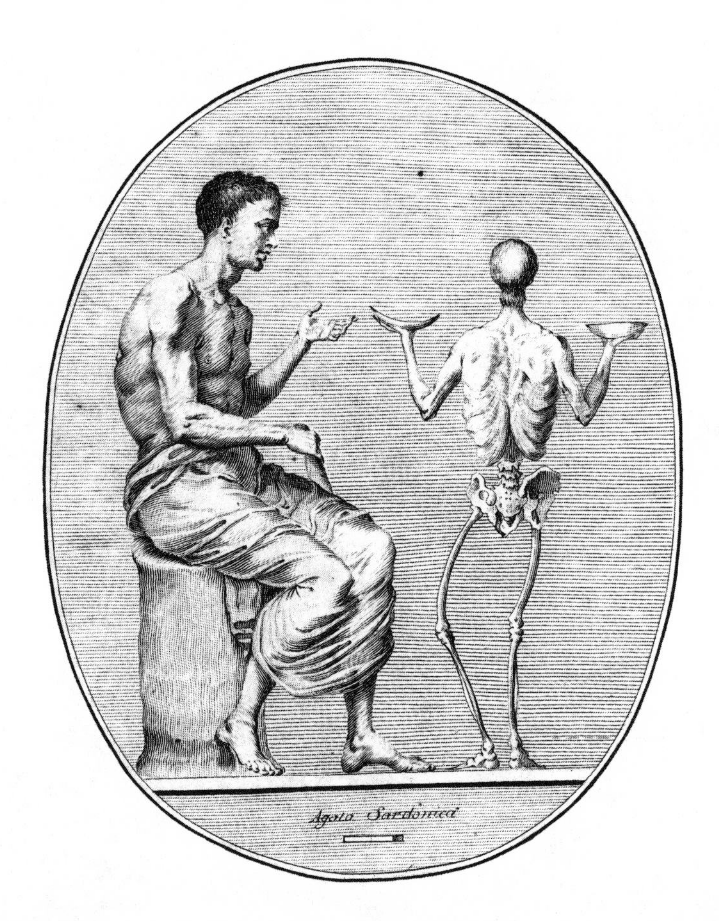

FIG. 6.6. Prometheus making the first man, half-completed with torso molded onto the skeleton. Engraving, *Alcuni monumenti del Museo Carrafa* (Naples, 1778), plate 25. Courtesy of Getty Research Institute, Los Angeles (89-B17579).

In the second gem type, Prometheus typically is shown affixing the arm bones to a human skeleton, as in figures 6.7–6.11. In figures 6.8 and 6.11 (plates 10 and 11), Prometheus uses a mallet or hammer to attach the arm to the skeleton.[29] In these images, the supposition is that he will then attach sinews and muscles to the framework of bones, adding internal organs, blood vessels, skin, hair, and so on—working outward from naturalistic interior anatomy to the finished human prototype.

In the context of the construction of a human form from internal anatomy to external features, it is illuminating to compare an ancient Chinese tale of artificial life. In this case a lifelike automaton was created

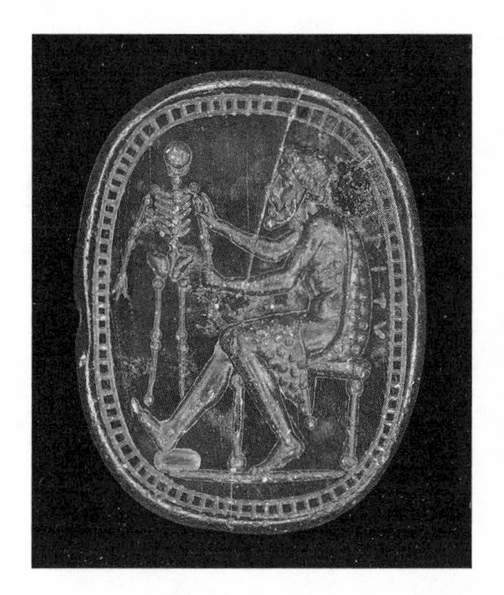

FIG. 6.7. Prometheus, seated, attaching the arm bone to the skeleton of the first human. Etruscan-style carved scarab (hatched border), inscription PIPITU, and cast, third to second century BC?, Townley Collection, inv. 1814,0704.1312. © The Trustees of the British Museum.

FIG. 6.8 (PLATE 10). Prometheus, seated, constructing the first human skeleton, using a mallet to attach the arm bone to the shoulder. Carnelian intaglio gem, date unknown, perhaps Townley Collection, inv. 1987,0212.250. © The Trustees of the British Museum.

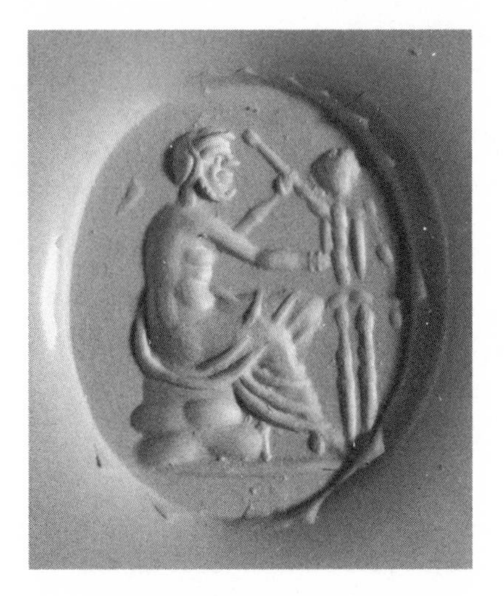

FIG. 6.9. Prometheus sitting on a rock, attaching raised arm to skeleton of first human, cast of carved gem, dark green jasper, first century BC, 82.AN.162.69. Courtesy of the Getty Museum.

FIG. 6.10. Prometheus attaching the arm to a skeleton, carnelian scarab, about 100 BC (modern gold ring setting). Boston Museum of Fine Arts, 62.184, Gift of Mrs. Harry Lyman.

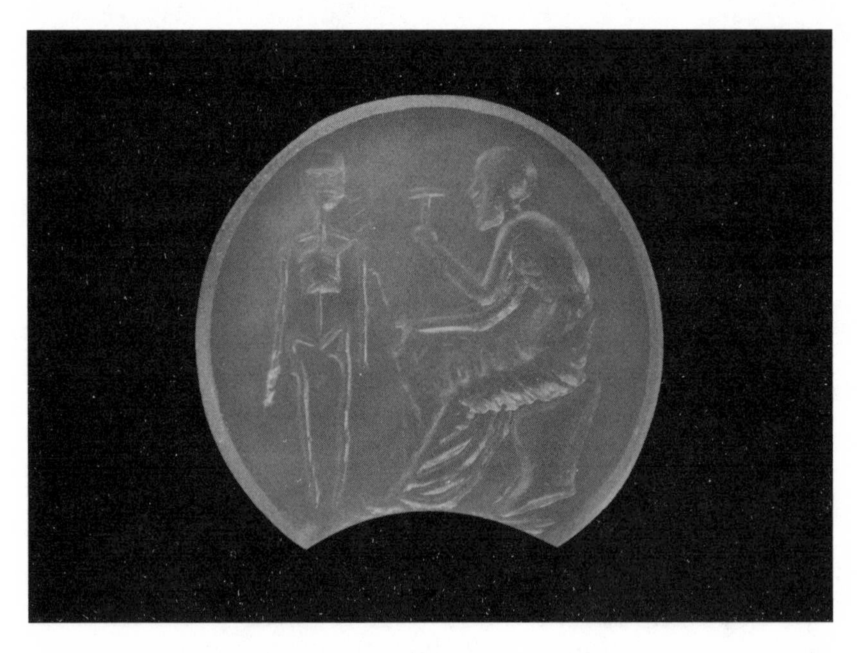

FIG. 6.11 (PLATE 11). Prometheus using a mallet to make a skeleton, chalcedony gem, first century BC, Thorvaldsens Museum, Denmark, acc. no. 185.

from the inside out with realistic and functional internal structure. Set during the reign of King Mu (ca. 976–922 BC) of the Zhou dynasty, the tale describes an android created by a master "artificer" named Yan (Yen Shih). The story appears in the *Book of Liezi*, attributed to the Daoist philosopher Lie Yukou (ca. 400 BC), although fixing the exact date is complex. In the tale, Master Yan introduces King Mu and his concubines to his marvelous man-made man, who walks, dances, sings, and otherwise perfectly mimics the actions of a real human being. The king is entranced—until the man flirts with the royal concubines. The king flies into a rage, then is astounded when Yan opens up the automaton to reveal its biotechnological construction, the "exact replication of human physiology in artificial form (*jiawu*)." Lifelike down to the finest detail, the outer body is made of leather, wood, hair, teeth, glue, and lacquer, and inside are artificial muscles and a jointed skeleton, with organs, liver, heart, lungs, intestines, spleen, kidneys—each of which controls specific bodily functions in Master Yan's android.

The ancient theme of building hyperrealistic androids from the inside out, beginning with anatomically exact skeletons and internal organs, evident in the Prometheus gems and in this Chinese tale, recurs in modern science fiction. For example, in the film *Blade Runner 2049*, the discovery of the buried skeletal remains of the runaway replicant Rachael reveals that replicants have "human" physiology—and might even be able to give birth to offspring.[30]

• • •

The artistic decision to show Prometheus constructing the first human starting with the bone structure likens the Titan to a sculptor who constructs a statue upon a model skeleton. *Kanaboi*, skeletal forms, usually of wood, were used by ancient sculptors as the internal core around which they attached clay, wax, or plaster in the first stages of creating statues. Wooden cores were also used with cold-hammered sheets of metal and in the lost-wax casting of bronze statues, as described in the writings of Pausanias, Pollux, Hesychius, and Photius. The artistic process is also mentioned by Pliny (34.18.45–47), who admired the excellently wrought small clay models and wooden skeletons used in the first stages of making bronze statues in the studio of the renowned sculptor Zenodorus in Rome. Wooden armatures would not survive the heat of casting, but modern analyses of famous ancient bronze statues reveal that metal armatures were also used. A *kanabos* served as a kind of three-dimensional diagram of body structure.[31] The scenes on the unusual gems discussed above show Prometheus designing his project, using technology and tools, and starting by assembling a real *kanabos*, the physical structure of what will become the first man.

In his treatises on biological anatomy and movement, Aristotle refers to *kanaboi*. He compares the way the network of blood vessels "displays the shape of the entire body . . . like the wooden skeleton (*kanabos*) used in artist's modeling." Moreover, Aristotle invokes familiar devices of his day, mechanical dolls or some sort of self-moving automata, as analogies to help explain the inner mechanical composition and workings of animals and humans. Referring to the skeleton as the framework that allows movement, Aristotle's language is mechanistic: he notes that animals have sinews and bones that function much like the cables attached to pegs or iron rods inside automata.[32]

The artistic representations of Prometheus working with sections of the human body and assembling a skeleton *kanabos* suggest that artists and viewers would understand his creation as a form of *biotechne*, analogous to a sculptor beginning with the interior framework to make automata that would then become the original *living* humans. In the first stage, he builds what viewers recognize as their own anatomy, logically assembling the progenitors of the human race from the inside out.

● ● ●

In all the variants of the Prometheus creation myth, the realistic forms of humans become the reality they portray: they become real men and women. This paradoxical perspective taps into the timeless idea that humans are somehow automata of the gods. The almost subconscious fear that we could be soulless machines manipulated by other powers poses a profound philosophical conundrum that has been pondered since ancient times: If we are the creations of the gods or unknown forces, how can we have self-identity, agency, and free will? Plato (*Laws* 644d–e) was one of the first to consider the idea of humans as nonautonomous: "Let us suppose that each of us living creatures is an ingenious puppet of the gods." The myth of the artificial woman Pandora, fabricated by the god Hephaestus, calls up similar questions, as we will see in chapter 8. These concerns about autonomy and soul also suffuse traditional Hindu, Buddhist, and Daoist tales about robots (above and chapter 5). In one Hindu story, for example, an entire city is populated by silent but animated townspeople and animals, later revealed to be realistic wooden puppets, all controlled by a solitary man on a throne in the palace.[33]

The notion that humans arose as the automata or playthings of an imperfect and/or evil demiurge and the ensuing questions of volition and morality were forcefully articulated in the ancient movement of Gnosticism (first through third century AD). In modern times, questions of human autonomy were debated by T. H. Huxley and William James in the 1800s, and Gnostic concepts are powerfully revived by philosopher John Gray in *Soul of a Marionette* (2015) and novelist Philip Pullman in the epic trilogy *His Dark Materials* (1995–2000). The *Blade Runner* films (1982, 2017) are another example of how science-fiction narratives play on the paranoid suspicion that our world is already full of androids—and that it would be impossible to apply a Turing test to oneself to prove that one is not an android.[34]

One of the replicants in *Blade Runner* repeats, "I think, therefore I am," the famous conclusion by the French philosopher René Descartes (1596–1650). Descartes was quite familiar with mechanical automata of his era powered by gears and springs, and he embraced the idea that the body is a machine. Anticipating Turing and similar tests, Descartes predicted that one day we might need a way to determine whether something was a machine or human. "If there were machines in the image of our bodies and capable of imitating our actions," wrote Descartes, then perhaps tests based on flexibility of behavior and linguistic abilities would expose nonhuman things.[35]

• • •

In the myth of Prometheus and Epimetheus, related by Plato (chapter 4), earth's creatures are created and then "programmed" with capabilities and defenses so that they will not fall into mutual destruction but will maintain equilibrium in nature. But the limits of biotechnology are revealed when the animals receive all the "apps" and nothing is left over for the humans, naked and defenseless. Feeling pity, Prometheus gives mortals craft and fire. Ever after, the Greek myths demonstrate how the immortal gods and goddesses play out their own power games, manipulating, withholding, rewarding, and punishing generations of mortals, for eternity. And soon enough, humankind itself would develop the urge to create and control life, like the gods. Many ages ago, the vision of capricious gods or careless, even evil, demiurges haphazardly doling out natural capabilities, and controlling or neglecting their human toys, sketched the outlines of one of the most chilling genres of science fiction still capturing audiences today.[36]

By the fifth century BC, the Athenians were venerating the rebel Prometheus and the precious gifts of technology he gave to humanity. The Titan was worshipped at an altar in what became the grove of Plato's Academy—alongside Athena and Hephaestus. During the city's most important civic festival, the Panathenaia, the Fire-Bringer Prometheus was honored with a relay torch race. Runners began at the altar in the Academy outside the city walls and wound through the Kerameikos, the district of potters and other craftspeople who revered Prometheus as their patron (along with Daedalus). The torch race culminated with the last runner kindling the sacred fire on Athena's altar on the Acropolis. A relief sculpture of Prometheus (and Hephaestus's creation, Pandora) decorated the base of the majestic statue of Athena in the Parthenon.[37]

• • •

In the Middle Ages and the Renaissance, Prometheus's theft of fire and his subsequent torment were transformed into an allegory for the human soul seeking enlightenment. Ever since, Prometheus has inspired artists, writers, thinkers, and scientists, as a symbol of creativity, inventive genius, humanism, reason, and heroic endurance and resistance against tyranny.[38]

Two famous literary works show how later authors were inspired by Prometheus's creations. In Shakespeare's *Othello* (1603), Othello says he cannot restore "Promethean heat" to Desdemona's dead body once her "light" is extinguished. The allusion refers to the notion that Prometheus himself bestowed life on his clay figures with the fire he stole from the heavens.

"Promethean heat," in the form of electricity, animates the monster created from grafted parts of pillaged corpses in the sensational scene in the iconic 1931 film *Frankenstein* starring Boris Karloff, which was based on the celebrated novel *Frankenstein* by Mary Shelley. Written in 1816 and published in 1818, Shelley's story was strongly shaped by classical mythology. Her father, William Godwin, wrote a commentary on seekers of artificial life in antiquity, including the witches Medea and Erichtho and the artisans Daedalus and Prometheus. Mary's companions Percy Shelley and Lord Byron were writing poems about Prometheus at the time. In the novel, Mary Shelley conceived of her scientific genius Victor Frankenstein as a Promethean "fire-bringer" for her era. She also drew on exciting scientific and pseudoscientific ideas about alchemy, occult transference of souls, chemistry, electricity, and human physiology current in her day.[39]

Some scholars suggest that Mary Shelley was influenced by reports of macabre dissection experiments carried out by the notorious alchemist Johann Dippel (b. 1673) of Frankenstein Castle, near the villa on Lake Geneva where she wrote the story. Debates over the electrostimulation work of Luigi Galvani and others were also much in the public eye by the 1790s. Shelley was certainly aware of morbid experiments in which animal and human corpses were grotesquely "reanimated" with electricity. A public demonstration of *galvanism* on the twitching cadaver of an executed criminal, for example, was staged in London in 1803. The life-giving principle was left vague in her 1818 novel, but Shelley does mention galvanism in her revised 1831 edition. She drew her subtitle, *The Modern Prometheus*, from the philosopher Immanuel Kant's famous essay (1756) warning about the overweening "unbridled curiosity" exemplified by Benjamin Franklin's "discovery" of electricity.[40]

Shelley tells how the young scientist Victor Frankenstein devotes two years of painstaking work to building an artificial, intelligent android. He assembles the creature part by part using raw materials from slaughterhouses and medical dissections. In light of Shelley's story of a "modern Prometheus," the ancient Etruscan illustrations, on gems, of Prometheus putting together human body parts and skeletons seem to take on an eerie prescience. In fact, the engravings of the Carafa gems in figures 6.5 and 6.6 were published in 1778. Several of the intaglios showing Prometheus working on the unfinished torsos and assembling skeletons were included in the vast collection of ancient and neoclassical gems amassed by the Scottish engraver and antiquarian James Tassie (1735–99). An illustrated two-volume catalogue of Tassie's collection was published in 1791.[41] Shelley and her circle may well have observed or heard described a number of gems featuring Prometheus making a human with body parts.

Yet another classical influence on Shelley's *Frankenstein* could have been the horrifying Thessalian necromancer Erichtho. A witch who haunts battlefields and graveyards seeking body parts for her spells, Erichtho most famously appears in Lucan's writings of the first century AD, a Latin poet well known to Shelley. In his *Civil War*, Lucan describes Erichtho striding grimly across a smoking battleground, seeking serviceable cadavers with intact lungs to resurrect. In a grisly scene, Erichtho uses dead animal parts to reanimate the human corpses. In imagery reminiscent of the witch Medea in Greek myths (chapters 1 and 2), Erichtho mutters incantations and gnashes her teeth as she compels the dead to come alive. The corpses jerk back to life convulsively, then walk about "remarkably quickly but stiff-limbed," evoking the stereotypical stiff-jointed walk of zombies, animated statues, and robots. Appalled to be unnaturally summoned back to life by the witch, the living dead throw themselves onto burning pyres.[42]

In Shelley's story, often hailed as the first modern science-fiction novel, the scientist hopes to create a humanoid of sublime beauty and soul. But the resulting creature is a hideous, sentient monster who wreaks havoc and bitterly resents being brought into existence. Some early modern thinkers saw the ancient myth of Prometheus's endless torture as a symbol of his gnawing doubts about his creation of humankind. Echoing Kant, some historians of robotics see the Promethean tale as a warning that anyone who "tries to build life artificially is acting outside the legitimate human province, carelessly straying into the divine orbit."[43] As in so many ancient myths and popular legends about artificial life achieved through mysterious supertechnology, Shelley's horror tale is a gripping meditation on the themes of striving to surpass human limits and the perils of scientific overreaching without full knowledge or understanding of the practical and ethical consequences.

• • •

In some accounts, Zeus asked Prometheus to make the first humans. But Zeus also meted out revenge on Prometheus for stealing fire and other tools to give to humans. (Zeus devised a separate eternal penalty for humanity, as well, as we shall see in the next chapter.) Ancient estimates of how long humanity's champion endured the torment of Zeus's Eagle range from thirty to one thousand to thirty thousand years. According to one strand of the myth, illustrated by many ancient artists, at last Zeus gave Heracles permission to kill his huge *Aetos Kaukasios* ("Eagle of the Caucasus"), thus ending Prometheus's anguish.[44]

The divine torture-eagle had various origins, recounted in different versions of the myth. Of particular interest is the summary given by Hyginus, a Roman librarian (b. 64 BC) who compiled a wealth of mythological material from numerous Greek and Latin sources (many now lost) in two treatises, *Fabulae* and *Astronomica*. Reviewing the ancient traditions, Hyginus (*Astronomica* 2.15) reported, "Some have said that this eagle was born from Typhon and Echidna, others from Gaia and Tartarus, but many point out that the eagle was made by the hands of Hephaestus." This tradition mentioned by Hyginus, that the giant Eagle sent to ravage Prometheus was fashioned by the god of the forge, conjures an image of a kind of metallic drone-eagle set to home in on Prometheus's liver at a certain time each day.

Notably, Apollonius (*Argonautica* 2.1242–61) penned an extraordinary description of Zeus's great Eagle as an unnatural, gleaming bird of prey with machinelike movements. Jason and the Argonauts observe the "shining Eagle" returning to the Caucasus crag "each afternoon flying high above the ship with a strident whirr. It was near the clouds, yet it caused all their canvas sails to quiver to the beat of its wings. For its form was not that of an ordinary bird: the long quill-feathers of each wing rose and fell like a bank of polished oars."

There are several pieces of ancient literary evidence for the idea of metallic birds of prey. The man-eating Stymphalian Birds, for example, were destroyed by Heracles in his Sixth Labor. The monster birds were often visualized with bronze feathers and armor-piercing beaks. From central Asian epic comes another image of robotic raptors. In the folk traditions about Gesar of Ling, the evil hermit Ratna makes and dispatches a trio of sinister giant metal birds to kill the hero Gesar. With rattling feathers that are "thin blades of iron and copper" and "beaks like swords," the birds swoop down on young Gesar, who fells them with three arrows.[45]

Mechanical birds were actually constructed as early as the fifth and fourth centuries BC in Greece. There was a bronze eagle that flew up to signal the start of the horse races at the Olympic Games (described by Pausanias 6.20.12–14) and a flying dove model was attributed to the scientist Archytas. As noted in chapter 1, Apollonius would have observed numerous automata and self-moving devices in Ptolemaic Alexandria (see chapter 9 for these and other historical inventions).[46]

Zeus's Eagle, fabricated by the god Hephaestus, would not be the only artificial animal created expressly as a killing or torture device in Greek myth and history, as the following chapters reveal. Throngs of animated devices and creatures "made, not born" fill out Hephaestus's stellar résumé of ingenious artifices and automated devices. Some are laborsaving, but others are deliberately intended to inflict harm.

CHAPTER 7

HEPHAESTUS

DIVINE DEVICES AND AUTOMATA

ONLY ONE GOD in Greco-Roman mythology has a trade. Not only does this god engage in strenuous physical labor; he even breaks a sweat. This same god possesses great intelligence, and his technological productions evoke universal wonder. The hardworking god is Hephaestus, supreme master of metalworking, craftsmanship, and invention.

An outsider among the other divinities, the blacksmith Hephaestus was lame and by some accounts had no father. Both his mother, Hera, and his wife, Aphrodite, rejected him; he was even cast out of Mount Olympus for a time. Yet all the gods and goddesses were in awe of Hephaestus. They called on the smith god whenever they required something of beautiful or clever design and sublime craftsmanship. Hephaestus created the divinities' gold and marble palaces secured with unbreakable locks. He made special weapons, armor, and equipment for gods and heroes: a partial list includes arrows for Apollo and Artemis; the Medusa shield for the hero Peleus; armor for Heracles, Achilles, Diomedes, and Memnon; Athena's spear and Apollo's chariot. He made an ivory replacement shoulder blade for the hero Pelops. For King Aeetes, Medea's father, he made the fire-snorting bronze bulls, and he engineered four fabulous fountains that provided wine, milk, oil, and hot and cold water. Against his will, Hephaestus was ordered by Zeus to make the chains that shackled Prometheus on the mountain, and the smith god forged Zeus's dread lightning bolts, depicted in art as a stylized bundle of metal projectiles hurled like a javelin. Zeus's scepter was another of his works—this was said to have been given to the mythical King Agamemnon of Trojan War fame. The scepter was displayed in a temple in Chaeronea, one of the several artifacts attributed to Hephaestus seen by Pausanias (9.40.11–12).[1]

The earliest description of Hephaestus at his forge appears in an extended passage in the *Iliad*. In the scene, the goddess Thetis seeks out Hephaestus to create a glorious set of armor for her son, Achilles (fig. 7.1). She finds the smith "glazed with sweat," working at his anvil in his abode made of bronze, where he is aided by various automated devices. Hephaestus wipes his brow with a sponge, sets aside his project, stores his tools in a silver chest, and greets his guest.

Thetis requests a bronze helmet, a richly decorated shield, and chest and leg armor more fabulous than any other ever made. Elaborate

FIG. 7.1. Hephaestus in his forge, showing Thetis the marvelous armor for her son, Achilles. Red-figure kylix, from Vulci, about 490–490 BC, by the Foundry Painter, F 2294. Bpk Bildagentur / Photo by Johannes Laurentius / Antikensammlung, Staatliche Museen, Berlin / Art Resource, NY.

descriptions of the individual pieces of armor follow. The shield is the centerpiece, made of "fine bronze, tin, silver, and gold" and "forged in five layers" with a "triple-ply rim." Homer's detailed description of the sophisticated technology of the shield's construction attracts the attention of modern engineers, such as Stepfanos Paipetis. Paipetis notes that Hephaestus uses composite materials to make "successive metal laminates with very different properties." The god's craftsmanship represents the ideal perfection of a human smith's knowledge of "dynamic mechanical properties of laminated composite structures," either observed in Homer's own day (eighth century BC) or perhaps transmitted from earlier times in oral traditions.[2]

Later in the *Iliad*, on the battlefield at Troy, Achilles and his companions admire the magnificent armor intricately embossed with dazzling panoramas that seem alive. The scenes on the divinely wrought shield reflect a marvelous "artificial world complete with motion, sound, and lifelike figures."[3] As if in a "movie in animated metal," the people on the shield's scenes are "vigorous and moving; they can sense, reason, and argue," and they have voices, "like living mortals." Homer's description is reminiscent of the eerily true-to-life images that frightened Odysseus in the Underworld and prefigures the "virtual reality" *phantasia* productions by the artist Theon of Samos (fourth century BC), which incorporated sounds, music, and lights (chapter 5). In the curious and paradoxical *Iliad* passage, Homer stresses the astounding realism of the scenes on the shield, specifying the different metals and techniques that Hephaestus used to "construct the various figures" while "calling attention to their crafted realism." The description causes one to wonder, "Could this verbal description have achieved any of this precision without referencing some visual artifact?"[4]

• • •

Before we move on to Hephaestus's other marvels and his artificial life projects, it is worth pausing to recognize that metal armor was one of the earliest artificial human enhancements (chapter 4). Bronze armor was designed to make warriors' bodies less vulnerable. But what is most striking about the bronze armor of classical antiquity is its form. The main piece of armor, the cuirass or chest plate, was molded to look like an idealized

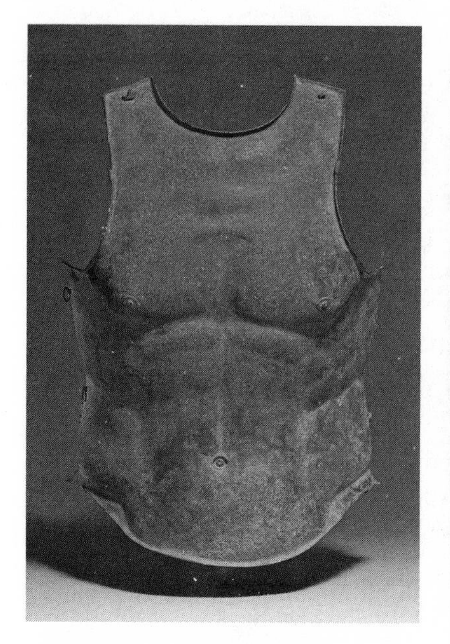

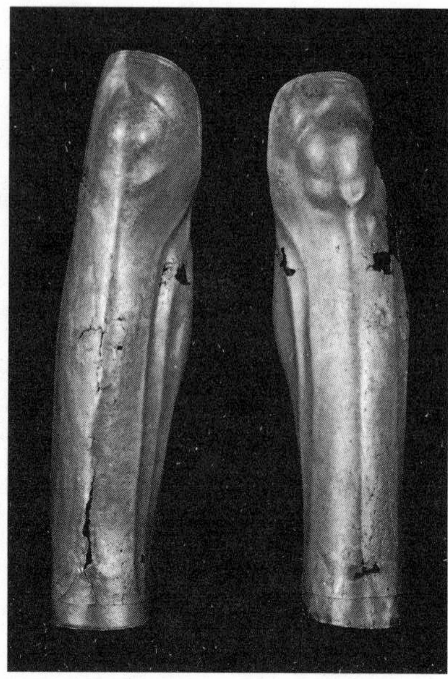

FIG. 7.2. Muscle cuirass, bronze, Greek, fourth century BC, 92.180.3 © The Metropolitan Museum, Art Resource, NY. Greaves, realistic leg armor, fourth century BC, Archaeological Museum, Sofia, Bulgaria. Erich Lessing / Art Resource, NY.

male physique cast in bronze. The "anatomical" armor, also called the "heroic" or "muscle" cuirass, first appeared in archaic Greece and became widespread by the fifth century BC. It was cast in two pieces, front and back, attached by straps. The hammered bronze cuirass was made to fit a man's upper body, with realistic details in relief to mimic the bare torso of a "hero," with nipples, navel, and impressively sculpted pectoral and abdominal muscles, resembling those of the mythic strongman Heracles. The greaves, bronze shin guards, were also shaped to delineate the knee and calf muscles.

A Greek hoplite who donned the artificial human enhancement of bronze chest and leg armor was essentially donning an exoskeleton that replicated the outer appearance of an idealized, "heroically nude" bronze statue. Notably, the heroic bronze cuirass worn by ordinary Greek soldiers on ancient vase paintings (fig. 7.3) resembles the robust bronze body of the automaton Talos, painted yellowish white (compare figs. 1.3, 1.4, plate 1). The bronze chest plate and greaves transformed every soldier—no

FIG. 7.3. Vase painting of “heroic” cuirass, 325 BC, National Archaeological Museum of Spain. Photo by Marie-Lan Nguyen.

matter what his body type—into a formidable, muscle-bound warrior. An advancing, clanking phalanx of Greek hoplite soldiers clad in muscle armor would present a living wall of superhuman bronze warriors.[5]

Later, the Romans took up the heroic cuirass molded to look Herculean. The Romans further embellished the ceremonial armor and sometimes included realistic silver face masks, which resulted in the appearance of a fully metallic superwarrior. Other military cultures fashioned armor intended to frighten enemies with the semblance of an army of iron men, for example, the eerie iron face masks of the Kipchak of central Asia (see chapter 4 for a medieval Islamic tale about Alexander's iron cavalry). By the Middle Ages in Europe, full body armor as a metal exoskeleton had evolved into elaborate, heavy suits of armor, as knights dueled with swords and jousted with spears. As we saw in chapter 1, today's military scientists are reviving a highly advanced exoskeleton idea, modeled on the mythic figure of Talos, to be further enhanced with computers and sensors.

• • •

As a god, Hephaestus was capable of workmanship and engineering superior to what could be achieved by mortal artisans. His works displayed prodigious creativity and skills, surpassing those of his earthly parallel, the legendary Daedalus. But like Daedalus and the Titan Prometheus, Hephaestus was imagined using implements and methods resembling those used by real smiths and artisans. And like Daedalus and other craftsmen, in ancient art and literature Hephaestus was portrayed at work surrounded by his tools and half-completed devices and statues. Generic scenes of smiths and sculptors at work closely mirrored the typical scenes of Hephaestus at work in his forge, in Greek vase paintings and in Roman frescoes (Hephaestus was called Vulcan in Rome).[6]

Many of the items of Hephaestus's manufacture were made expressly for gods and goddesses. To enable the divinities to drive their chariots with ease in and out of their Olympian abode, for example, he made gates that swiveled "on their hinges of their own accord, *automatai*"—thus, jokes classicist Daniel Mendelsohn, "anticipating by nearly thirty centuries the automatic garage door."[7]

Two cunning devices were wielded against Hephaestus's unfaithful wife, Aphrodite, and his uncaring mother, Hera. In one myth, Hephaestus

FIG. 7.4 (PLATE 4). Blacksmith at work, with tools, red-figure kylix, late sixth century BC, 1980.7. Bpk Bildagentur/ Photo by Johannes Laurentius / Antikensammlung, Staatiche Museen, Berlin / Art Resource, NY.

fashioned a nearly invisible net of incredibly fine but strong metallic mesh to ensnare Aphrodite in bed with the war god Ares. To take revenge on Hera for rejecting him, Hephaestus presented his mother with a golden throne cleverly devised to include a trap set with some mechanism, perhaps a spring or lever, to restrain her as soon as she sat down. Hera was stuck until Hephaestus released her. The scene of Hera on the throne is depicted on several ancient vase paintings. In one, Hephaestus is shown actually releasing the fetters.[8]

Hera, lacking her son's technology, deployed a supernatural creature named Argus as a sentinel against her husband, Zeus. Argus's special powers could be seen as a form of divine artificial enhancement. In a fragment of a Hesiod poem, *Aegimius*, and subsequent texts, Argus was a giant guardian sent by Hera to defend the nymph Io when she was in the form of a heifer being pursued by Zeus. Called *Panoptes* ("all-seeing"),

FIG. 7.5. Top, blacksmith tools, about 250 BC, Museum für Vorgeschichte, Asparn, Zaya, Austria. Erich Lessing / Art Resource, NY. Bottom, ancient blacksmith tools, from the Byci Skala cave, Czech Republic, sixth–fifth century BC, Naturhistorisches Museum, Vienna. Erich Lessing / Art Resource, NY.

Argus never slept and could see in all directions with his many eyes, ranging from four to a hundred depending on the source. On vases painted in the sixth to fourth century BC, the body of Argus Panoptes is shown entirely covered with eyes, as described by the mythographer Apollodorus. A fine wine jug (*lekythos*) of 470 BC by the Pan Painter was recently discovered in ancient Aphytis, northern Greece (fig. 7.6). The body of the humanoid Argus is covered in eyes and has a janiform head looking in opposite directions.[9]

FIG. 7.6. Argus with many eyes and janiform head. Attic red-figure lekythos from Aphytis, by the Pan Painter, about 470 BC. © Hellenic Ministry of Culture and Sports, courtesy of Ephorate of Antiquities of Chalcidice and Mount Athos.

The ancient myth of a hypervigilant watcher that never sleeps and observes from all angles inspired Jeremy Bentham's eighteenth-century panopticon designs for institutions and prisons, heralding the proliferation of banks of surveillance cameras ubiquitous in the modern world. Accordingly, numerous security providers operate under the name "Argos/Argus." The computerized exoskeleton TALOS suit to augment soldiers' senses to be developed by US military scientists also features multiple "eyes" like Argus's (chapter 1), while other military scientists seek ways to create soldiers who can forgo sleep, like Hera's sentinel (chapter 4).[10]

• • •

The most captivating devices created by Hephaestus were those described as exceedingly lifelike and/or as self-moving automata that mimicked natural bodily forms and possessed something like mind. We have already met some of Hephaestus's artificial animated creatures: the bronze guardian Talos of Crete, the *Khalkotauroi*, fire-breathing bronze bulls wrangled by Jason, and Zeus's torturing Eagle. Other lifelike animals made by Hephaestus include horses, dogs, and a lion. Except for Talos, the animating mechanisms or inner workings of these metallic wonders are not described in any surviving texts.[11] But it is telling that they are made by the inventor god, the same god who forged Talos and other automata via *techne*.

Most of the accounts of Hephaestus's animal-shaped devices are very ancient. An exception is a story by the late Byzantine-era epic poet Nonnus (*Dionysiaca* 29.193), who imagined Hephaestus creating a pair of animated bronze horses to draw the adamantine chariot of his sons, the Cabeiroi. As with the brazen bulls, flames shoot from the horses' mouths. "Their bronze hooves beat the dust with a rattling sound," and the equine automata even emit a "dry whinnying sound from their throats." By the time of Nonnus, the fifth century AD, a number of inventors had been building actual self-moving devices for several centuries (chapter 9). Some of these real creations may have inspired Nonnus's vision of the flame-snorting horses as a kind of poetic double of the ancient myth of the bronze bulls.

Much earlier—but puzzling—artistic evidence for a horse made by Hephaestus appears on a unique Etruscan mirror made in the fourth century BC. The horse statue and inscriptions engraved on the bronze mirror have stumped Etruscan scholars and classical art historians. The Etruscans, as we know, told their own oral versions of Greek mythology. The scene on the mirror shows a realistic metal horse statue (labeled *Pecse*) being created by *Sethlans*, the Etruscan Hephaestus, and an assistant named *Etule* wielding a smith's hammer (fig. 7.7, plate 8).

The horse labeled *Pecse* has been identified by some scholars as the Trojan Horse, but questions arise with that interpretation. *Pecse* is the Etruscan name for Pegasus, but the horse on the mirror has no wings, and in the Greek myth Pegasus was born from the Gorgon's decapitated head, not forged by Hephaestus. This horse has no wheels; the Trojan Horse is wheeled in the earliest Greek artistic images.[12] No known Greek myths associate Hephaestus with the Trojan Horse. According to Homer (*Odyssey* 8.493), the Trojan Horse was constructed of wood by a Greek craftsman named Epeius, not by Hephaestus, and it was either made with Athena's help or else dedicated to Athena (see fig. 5.4, for this scenario on an Athenian vase by the Foundry Painter).

Who is *Etule*? It is possible that Etule is meant to be Epeius, but if this is an Etruscan version of the Trojan Horse story, he was inspired or guided by Hephaestus, instead of Athena. Epeius did have an Italian association: he was the mythic founder of the Greek colony Metapontum (in southern Italy), and it was said that the citizens displayed his tools in the temple to Athena there.[13]

On the Etruscan mirror, Sethlans/Hephaestus is doing something with some lumpy material around the horse's neck. In his right hand he is holding some of the same material. He appears to be removing or applying clay or making a plaster mold, like those used in ancient bronze casting techniques. A comparable scene appears on an earlier red-figure Athenian vase painting of about 460 BC. This vase has an unusual scene of a god other than Hephaestus actually working to make an artificially lifelike being. Figure 7.8 (plate 9) shows the goddess Athena, the patroness of Athenian craftsmen, making a clay model of a horse (the Trojan Horse). The hind leg is unfinished and its body is still rough. Behind Athena are tools like those used by Daedalus and Hephaestus and ordinary craftsmen in their workshops: a saw, drill, and bow drill. There is a mound of clay at her feet, and she is applying a handful of the clay to the horse's head. This classical vase image of Athena making a horse with

FIG. 7.7 (PLATE 8). Hephaestus (Sethlans) and assistant (Etule) making an artificial horse (Pecse), Etruscan bronze mirror, fourth century BC, from Orvieto, BnF Cabinet des Medailles, Bronze.1333, photo Serge Oboukhoff © BnF/CNRS-Maison Archéologie & Ethnologie, 2011. B. Woodcut of mirror, Victor Duruy, *History of Greece* (Boston, 1890), redrawn by Michele Angel.

FIG. 7.8 (PLATE 9). Athena making a clay model of a horse; she is holding a handful of clay and there is a pile of clay at her feet. Above left, a saw, drill, and bow drill. The horse's back leg is unfinished. Athenian red-figure wine jug, about 460 BC, F 2415. Bpk Bildagentur / Photo by Johannes Laurentius / Antikensammlung, Staatliche Museen, Berlin / Art Resource, NY.

clay is remarkably similar to the image of Sethlans/Hephaestus molding clay on the horse's neck on the Etruscan mirror.[14]

Looking more closely at the image on the Etruscan mirror (fig. 7.7, plate 8) one notices that the lively-looking artificial horse is chained by its front foot to a rock hobble. This is an odd detail for a lifeless statue. Odd, that is, until we recall the ancient Athenian jokes about needing to tether or bind "living statues" to prevent them from running away (chapter 5). The chain on the horse's leg could emphasize how realistic the artificial horse is—or it could indicate that Sethlans/Hephaestus and his assistant are making an animated statue of a horse, apparently illustrating an unknown Etruscan tradition.

• • •

Besides the bronze *phylax empsychos* ("animated guard") Talos, Hephaestus fashioned two other gifts for Minos. One was magical—a quiver full of arrows (or a javelin) that never missed their mark. The other item is more interesting: a supernaturally swift hunting dog that never lost its prey (the dog's image appears on the other side of coins of Crete depicting Talos). Sometimes viewed as an automaton hound, and sometimes as a wonder-dog with enhanced natural abilities, this mythic canine creation had many adventures. Often called Laelaps, the dog features in a story (part of a lost Homeric epic, the *Epigoni*) that begins with Minos.

His wife, the witch Pasiphae, we recall, had cursed Minos with scorpion ejaculations to keep him faithful (chapter 4). Minos is finally cured of that malady with a reverse spell cast by another witch, named Procris. Minos gives the special hound Laelaps to Procris in gratitude. Then Procris's husband, Cephalus, takes Laelaps to Boeotia, in Greece, to hunt the Teumessian Fox, a monstrous fox that could never be caught. This fantastical hunt sets up the sort of paradoxical conundrum that was so popular in Greek mythology and philosophy. The dilemma of a hound that cannot fail to catch prey and a fox that cannot be caught is resolved when Zeus transforms both hound and fox to stone. A pair of rock formations in the shape of the two animals was a famous ancient attraction near Thebes.[15]

Confusingly, the hound of Crete/Laelaps story is entangled with the myth of the Golden Hound. Rhea, Zeus's mother, set this animated hound made of gold to guard the infant Zeus when he was hidden on Crete from his murderous father, Cronus. Who made this golden watchdog? Some say the Golden Hound was made by the metalworking gnomes or daimons called Kouretes or Dactyloi, who were charged with protecting the infant Zeus on Crete. (They were associated with the Telchines, who made the fabled living statues of Rhodes; chapter 5). But other sources say the Golden Hound was made by Hephaestus. At any rate, when Zeus assumed power on Mount Olympus, he ordered the Golden Hound to continue to guard the sacred site of his infancy at his temple on Crete. According to one mythic thread, Pandareus stole this precious Golden Hound from Zeus's temple, but the god Hermes recovered the Hound for Zeus. The rescue of the Golden Hound was illustrated on an archaic vase painting of the early sixth century BC (fig. 7.9).

FIG. 7.9. The Golden Hound made by Hephaestus, recovered by Hermes, after it was stolen by Pandareas. Black-figure cup, about 575 BC, Heidelberg Painter, Louvre A478. © RMN-Grand Palais / Art Resource, NY.

In the second century BC, the poet Nicander of Colophon interwove threads of these various tales to praise the origins of the marvelously swift real-world Molossian and Chaonian hounds admired by Greek hunters: "They say these dogs are the descendants of a dog" that Hephaestus manufactured. Hephaestus, he wrote, "cast it in Demonesian bronze and set a soul (*psyche*) in it." This animated hound, recounts Nicander, was passed from Minos to Procris to Cephalus, and ultimately was turned to stone by Zeus. The poet's folklore phrase "they say" imagines that an animated dog of metal could copulate with a living dog and have offspring. Nicander plays with the idea that an artificial animal could be so "real" that it could even procreate, much as some later Roman-era writers pretended that Galatea and Pandora—neither one born of biological parents—were so "human" that they could reproduce. Nicander employs this poetic conceit to confer a divine pedigree on the best hunting hounds of antiquity, much as Athenian craftsmen claimed Daedalus as their ancestor (chapter 5).[16]

The earliest known story of animals wrought in metal by Hephaestus appears in Homer's *Odyssey* (7.91–98). The scene describes the pair of dogs, one silver and the other gold, that defended the splendid palace of the mythic king Alcinous of the Phaeacians, a mysterious advanced culture. Odysseus admires these ferocious watchdogs, "fashioned with cunning skill," standing guard at the richly decorated entrance gates.

Homer describes the ever-vigilant hounds as "deathless and ageless." Some interpret the myth to indicate that the mastiffs could move to attack and even bite intruders, but that is not clear and Homer does not say how. Another mythic tradition says these same gold and silver dogs had once helped the god Poseidon, who then gave them to Alcinous.[17]

Three versions of a previously unknown mythic tradition about a bronze lion constructed by Hephaestus to guard the island of Lesbos came to light in 1986. The accounts appear in a badly damaged fragment of papyrus from the second century AD. The earliest source in the fragment appears to be from the third century BC. According to the papyrus, this bronze lion was hidden on the coast of Lesbos to defend against attacks from mainland Anatolia. The story comports with the ancient and medieval belief that bronze statues could serve as guardians and "magic shields" (chapter 1), and some statues, like Talos and the Golden Hound, were further imagined as "animated" (*empsychos*).

The lion statue of Lesbos was made in a two-step process, recalling the "soul" placed in the bronze dog mentioned by Nicander. In this case, Hephaestus cast the hollow lion and then placed *pharmaka* (powerful substances) inside it. The "animating" *pharmaka* were "beneficial to mankind."[18] This process brings to mind Medea placing powerful *pharmaka* inside the hollow bronze statue of Artemis in chapter 2, and the internal life force inside Talos in the form of ichor (chapter 1). One might also note that the artificial lion "animated" by powers "beneficial to mankind" seems to anticipate the science-fiction author Isaac Asimov's first law of robotics (1942): A robot may not harm humans. That rule—broken by Talos and other ancient automata—still resonates with modern experts who work on the ethics of robotics and Artificial Intelligence. In the "23 Asilomar AI Principles" for ensuring ethical human values in Artificial Intelligence (set forth by the Future of Life Institute in 2017) the final rule states that "superintelligence should only be developed . . . for the benefit of all humanity."[19]

• • •

When the goddess Thetis interrupts him at his forge, Hephaestus is engaged in a project of "inspired artistry." Forging twenty bronze cauldrons on tripods mounted on golden wheels, he is in the act of riveting the handles, which have not yet been attached. Bronze tripods, three-legged stands for basins or cauldrons, were ubiquitous everyday furniture in classical antiquity. Ceremonial, ornate tripods were often dedicated in temples or presented as prizes and gifts. When completed, this very special fleet of tripods invented by Hephaestus could travel of their own accord, *automatoi*, delivering nectar and ambrosia to banquets of the gods and goddesses on command and then returning to Hephaestus (Homer *Iliad* 18.368–80). Unlike the ancient descriptions of Talos, no internal mechanism for the tripods was given by Homer, but they fit the definition of machines in that they can travel on their own and change direction.

The passages about the tripods and the automatically opening gates of Olympus (*Iliad* 5.749 and 18.376) are the earliest appearances of the ancient Greek word αὐτόματον, *automaton*, "acting of one's own will." In the fourth century BC, Aristotle quoted the Homeric verse and referred to the tripod-carts as *automata* (*Politics* 1.1253b). Notably, Philostratus (AD 170–245) reported that the peripatetic sage Apollonius of Tyana saw many amazing sights in India in the first or second century AD (*Life of Apollonius* 6.11). Among the *thaumata*, "wonders," were *tripodes de automatoi* and automated cupbearers that attended royal banquets. As many modern historians have remarked, the self-moving tripods serving the Olympian gods call to mind modern self-propelled, laborsaving machines, driverless cars, and military-industrial robots. Homer's myth reminds us that the impulse to "automate" is extremely ancient.[20]

Wheeled tripods do not appear in surviving ancient Greek art, and archaeological examples are unknown. However, many ornately decorated four-wheeled bronze carts for transporting cauldrons have been excavated in Mediterranean sites, dating to the Bronze Age (thirteenth to twelfth century BC). Today, one might speculate about tracks, springs, levers, strings, pulleys, weights, cranks, or magnets as plausible operating systems for self-moving tripods that behaved something like those in Homer's passage about Hephaestus. Indeed, a hypothetical working model of an automatic wheeled tripod can be viewed in the Kotsanas Museum of Ancient Greek Technology (near Pyrgos, Greece). The model uses millet grain, weights, ropes, and transverse pins, applying techniques developed by later historical engineers working in Alexandria, Philo and Heron (chapter 9).[21]

By the third century BC, Alexandria, Egypt, with its grand library and museum, had become a center for mechanical innovations. Perhaps inspired by Hephaestus's wheeled serving tripods in the *Iliad*, Philo (a Greek engineer born in Byzantium, but living in Alexandria) invented an automaton in the form of a woman who served wine. This robot was stationary but it could easily have been placed on wheels to move on an incline, using a simple design that would have been possible with materials, skills, and technology available in classical antiquity.[22] Just such a wheeled female servant automaton is described in the later Arabic treatise of AD 1206 by al-Jazari (b. AD 1136), a prolific practical engineer during Artuqid rule in eastern Asia Minor. In this design, liquid is poured into a vessel at the top and trickles into a basin until the basin tips and fills a cup in the servant's hand. The weight in the cup then causes the wheeled servant to roll down an inclined plane toward the drinker (many more historical self-moving devices and automata are discussed in chapter 9).[23]

The salient point about the self-driving tripods and similar fictions in Greek mythology about self-moving devices made by Hephaestus is that—in the time of Homer, more than twenty-five hundred years ago—ingeniously designed self-propelling carts manufactured by a super-smith were at least thinkable in the realm of mythology, even though the technology was not specified or known.[24]

Rolling tripods are absent in ancient Greek art, but there is a striking image of a *flying* tripod. It appears on a beautiful vase painting made in about 500–470 BC by the talented and prolific artist known as the Berlin Painter (fig. 7.10). The scene shows the god Apollo seated on a winged tripod flying over the sea above leaping dolphins. Everyone knew that the priestess of Apollo at the Delphic oracle sat on a special tripod while in a prophetic trance. A legend circulated in antiquity about a beautiful golden tripod, made by Hephaestus and owned by Helen of Troy, designated by the Delphic oracle for "the man most wise." According to the oracle, the tripod would travel on its own to the wisest man. The golden tripod passed among the Seven Sages and ultimately was dedicated to Apollo.[25] Could this curious legend be somehow related to the vase scene of Apollo's tripod "transformed into a fantastic flying machine"? The image is unique and the myth it illustrates is unknown.[26] Such a device would have been crafted by Hephaestus, who made the golden tripod, the special chair for his mother, and the fleet of self-propelled tripods to serve the gods. Indeed, plenty of

FIG. 7.10. Apollo seated on his tripod flying over the sea with dolphins and other marine creatures. Attic red-figure hydria, about 500–480 BC, Berlin Painter, Vatican Museums, Scala / Art Resource, NY.

literary and artistic evidence shows that the idea of flying "machines" in the form of wheeled chariots was current in archaic times.

Three of the many vase paintings depicting these flying chairs/chariot-cars are by the Berlin Painter, while the earliest known example is a vase of about 525 BC attributed to the Ambrosios Painter. The scene shows Hephaestus himself seated in a wheeled chair or chariot-car with wings, illustrating another unknown story (Hephaestus, we recall, was lame). Several other vases portray Triptolemus, associated with Demeter and the Eleusinian Mysteries, seated in or about to mount his flying wheeled chair-chariot (fig. 7.11). In this myth, the goddess sends Triptolemus to

FIG. 7.11. Triptolemus in his flying chair, with Kore, red-figure Attic cup found in Vulci, by the Aberdeen Painter, about 470 BC, Louvre G 452, Canino Collection, 1843, photo by Marie-Lan Nguyen, 2007.

disperse knowledge of agriculture over the earth, traveling in an airborne chair. Among the many ancient sources is a fragment of Sophocles's lost play about Triptolemus (468 BC) that describes him flying about in his special chair. Wings were not mentioned in the written sources—the wings were added later by vase artists as a way of indicating flight. We can guess that wings were attached to the flying machines of Apollo and Hephaestus for the same reason, to show that the wondrous vehicles were self-moving and capable of flight.[27]

* * *

The tripods created by the blacksmith god were mindless machines. But Hephaestus also fabricated wondrous automata in the shape of human beings with special abilities. One example appears in a fragment of a lost poem by Pindar. The scrap of poetry tells how Hephaestus made a bronze temple for Apollo, god of music, at Delphi. The pediment of the temple was graced by the *Keledones Chryseai*, "Golden Charmers," six golden statues of women who could sing. In the second century AD, the Greek traveler Pausanias (10.5.12) investigated the existence of the singing statues. He visited the site but learned that the bronze temple and the statues had long ago either toppled into a chasm during an earthquake or melted in a fire.[28]

Yet another group of automata wrought by Hephaestus represents a stunning "evolutionary leap forward" in replicating lifelike humanoids.[29] In the *Iliad* scene of the visit of Thetis to Hephaestus's forge, Thetis observes something astonishing: a staff of self-moving, *thinking* female automata who assist Hephaestus. These female assistants surpass the functionalities of the automatic gates, the traveling tripods, the singing statues on the roof at Delphi, and even Talos, the bronze guard who seemed to possess a kind of agency and consciousness. "Fashioned of gold in the image of maidens, the servants moved quickly, bustling around their master like living women" (*Iliad* 18.410–25). As the writer Philostratus remarked several centuries later (*Life of Apollonius* 6.11), "Hephaestus constructed handmaids of gold [and] made the gold breathe."

These humanoid helpers are not merely ultrarealistic "living statues" of gold with the ability to move, however. Hephaestus "built the mechanical serving girls" and then placed "within them mind, wits, voice, and vigor" (*noos, phrenes, aude, sthenos*) as well as the skills and knowledge of all the immortal gods.[30] So these golden assistants of Hephaestus are not only spontaneously mobile, but they anticipate and respond to his needs. And they are endowed with the hallmarks of human beings: consciousness, intelligence, learning, reason, and speech. (The people on Achilles's fabulous shield were endowed with the same capabilities, above.) "Hephaestus's Golden Maidens set the standard for artificial life," remarks a scholar of classical and modern science fiction. With "human intelligence and bodies indistinguishable from the real thing," the Golden Maidens are exceptional "divine artifacts in that they are composed of metal but have human-like abilities." The mythic gold helpers seem to presage modern notions of thought-controlled machines and AI. Like other automata made by Hephaestus, however, their inner workings are cryptic "black boxes."[31]

Yet the human-like qualities of the Golden Maidens could be seen as an ancient version of "Artificial Intelligence."[32] In effect, they are endowed with what AI specialists term "augmented intelligence," based on "big data" and "machine learning." In what might appear to be a case of mythic overkill, the *Iliad*'s female androids are described as a kind of storehouse of all divine knowledge.[33] In modern contexts, AI entities destined for specific tasks usually require no more information than would be needed for efficiency in problem solving. They need to be able to access useful knowledge but do not require a massive and indiscriminate "data dump." But just as it is difficult for modern AI developers to anticipate exactly what knowledge could be relevant to complex tasks or might become necessary down the road, the Homeric myth imagines that the gods would naturally wish to imbue Hephaestus's marvelous automata with a wealth of divine knowledge.[34]

• • •

The automata described in the *Iliad* are not the only self-moving entities in ancient literature imagined as possessing some form of intelligence and agency. In the *Argonautica*, for example, a supernatural oak beam in Jason's ship, the *Argo*, can speak and prophesy. Even more compelling in terms of an ancient vision of "Artificial Intelligence," however, are the remarkable ships of the Phaeacians, inhabitants of the technologically marvelous land encountered by Odysseus, in Homer's *Odyssey* (7–8). Phaeacian ships require no rudders or oars, no human pilots, navigators, or rowers, but are steered by thought alone. The Homeric myth envisions the vessels as controlled by some sort of centralized system, with access to a vast data archive of "virtual" maps and navigation charts of the entire ancient world. King Alcinous boasts that his unsinkable ships can travel very long distances under any weather and sea conditions and return to his port on the same day. The ships themselves "understand what we are thinking about and want," explains Alcinous, "They know all the cities and countries in the whole world and can traverse the sea even when it is clouded with mist, so there is no danger of being wrecked or coming to any harm." To transport Odysseus back to Ithaca, the ships simply "need to be told his city and country and they will devise the route accordingly." Odysseus marvels at the steady course of the pilotless Phaeacian ship, as swift as a falcon, as it carries him across the sea to his home island. The analogy to modern Global Positioning Systems (GPS) and automatic pilot and navigation systems is inescapable.[35]

Incidentally, a group of ancient Egyptian tales describe ships powered by artificially animated oarsmen. The texts are found in fragments of demotic papyrus pages dating to the Ptolemaic-Roman period (fourth century BC–fourth century AD). Set in the historical time of Ramses II, these stories tell how evil sorcerers make wax models of ships and rowers and command the figures to carry out tasks. It is interesting that the rowers are not only animated but apparently capable of independent thought and actions while completing their missions.[36]

● ● ●

Hephaestus's self-moving tripods and automated female servants have piqued the interest of historians of robotics. Their glamor overshadows yet another set of automated objects that have received less attention, although they too perform specialized labor in Hephaestus's forge.[37] Invented in antiquity to deliver more air to increase combustion and heat, real bellows technology was crucial in the development of metallurgy, which requires extremely hot fires. Later in the *Iliad* scene (18.468–74),

Hephaestus sets in motion twenty bellows that are self-operating and self-adjusting according to his needs. In the scene, Hephaestus "turns the bellows toward the fire and gives them their orders for working. The bellows begin to blow on the crucibles, blasting forced air from all directions wherever he required hotter or lower flames, following him as Hephaestus goes to and fro, working on his great anvil with his ponderous hammer and tongs." Like the automated doors of Olympus that open and close on their own, the traveling tripods, and the Golden Maidens, the bank of automatic bellows to stoke the blacksmith's fires were imaginary mechanical, laborsaving machines, doing work that would otherwise be done by living assistants or slaves.[38]

• • •

One of the essential motivations for the creation of machines and robots is economic. By performing mechanized labor, they relieve their masters of tedious toil. This line of thinking led Aristotle, in about 322 BC, to speculate about the socioeconomic implications of inventions like those described in Greek myths about automata (*Politics* 1.3–4). First, Aristotle compares human slaves to tools or automata that fulfill the wills of masters. To live well, he notes, one depends on "instruments, some of which are alive [and] others inanimate." Thus, for "the pilot of a ship, his tiller is without life [and] his sailor is alive." Aristotle continues, "A servant is like an instrument in many arts [and] a slave is an animated instrument—but a servant or a slave that can minister of himself is more valuable than any other instrument."

Aristotle's discussion is part of his defense of slavery. But then, in a remarkable passage, Aristotle engages in a thought experiment, suggesting a condition that might preclude slavery. If inanimate instruments could carry out their work themselves, he muses, then servitude might be abolished. "If every tool could perform its own work when ordered to do so or in anticipation of the need, like the statues of Daedalus or the tripods of Hephaestus, which the poet tells us could of their own accord move into the assembly of the gods," and "if in the same manner, shuttles could weave and picks could play *kitharas* (stringed lyres) by themselves, then craftsmen would have no need of servants and masters would have no need of slaves."[39]

Today, the ancient speculative fantasy that machines could free many workers from drudgery and replace slaves has become a commonplace reality in many parts of the world. Ironically, however, industrial robotics technologies now threaten to abolish human wage earners' livelihoods, leaving masses of idle, unpaid workers.

Meanwhile, dystopian science fictions paint nightmarish scenarios of a new, rising "servile class" of automaton-slaves that ultimately will rebel. The idea that creations of superior masters might revolt against their makers is also quite ancient. More than two millennia before Karel Čapek coined the word *robota* (derived from "slave"), the link between slavery and robots was already evident in Aristotle's passages, above, and in Socrates's comments about tethering living statues lest they escape and become useless to their masters, like runaway slaves (chapter 5). The theme is taken up in Jo Walton's percipient science-fiction trilogy set in classical antiquity, in which the goddess Athena establishes an experimental city based on Plato's *Republic*. Athena imports robots from the future to be mindless worker-slaves, but Socrates discovers that the robots not only possess consciousness but yearn for liberty.[40]

• • •

Modern historians of robotics and artificial life have so far only superficially addressed the question of whether or not the mythic moving statues of humans and animals, the driverless tripod-carts, the singing statues and mobile servants made by Hephaestus and other bronze workers should be considered mechanical automata. For example, Berryman maintains that Hephaestus's golden handmaids and the tripods could not have been imagined as products of "material technology" because "the technology of [Homer's] day" was not advanced enough to contemplate the idea of self-moving automata. "It may be tempting to read accounts of [ancient] 'statues that move' as anticipating modern robots," she remarks, but this is "not warranted, unless there is evidence of technology available" already that could make such things conceivable (Berryman's argument omits the bronze automaton Talos).[41] Truitt's history of medieval robots briefly discusses Hephaestus's tripods and golden assistants, but not Talos.[42] In his discussion of the four categories of automata in Greek mythology, Kang mentions the self-moving tripods, but leaves out the more relevant example of Hephaestus's female automata endowed with mind, strength, knowledge, and voice.[43]

The imaginary automata in question are, of course, located in mythical material, and their workings are not fully described in the extant ancient texts, but it is appropriate to consider how such entities were conceived of and visualized in ancient literature and art. Admittedly, the written material about mythic automata that survives from antiquity is incomplete and often contradictory. And the artistic evidence that exists today represents a minuscule portion of what existed in antiquity. Even so, it is worthwhile to glean as much information as one can about automata from Homeric times to the late Roman era, to try to understand all the ways that artificial life could be envisioned by ancient people. Any animal and human forms that were described as manufactured—that is, made, not born biologically—were products of what can be termed *biotechne*, life by craft, and therefore they deserve serious attention as the earliest *imaginings* of artificial life. Moreover, the many visualizations of artificial life in the mythic writings were put to good use in antiquity, as provocative ways to think about alternative worlds, which in turn raised ethical and philosophical questions about agency and slavery.

The surviving literary and artistic evidence, even though only a fraction of what once existed, shows that as early as the very first Greek writings in the time of Homer and Hesiod, people were already dreaming up notions of animated statues and self-moving contraptions. The myths demonstrate that automata were *thinkable*, long before technology made them feasible. Some, but not all, lifelike facsimiles were willed to come to life by mystical divine forces, like Pygmalion's ivory maiden. But as we have seen, many other self-moving "machines" and artificial beings were produced by inventors of myth and legend who were renowned for their technological prowess and ingenuity with clay and metal. The evidence demonstrates that nearly three thousand years ago people could express in mythological terms the idea that some type of exceptional technology might be capable of manipulating familiar materials, tools, and processes to make animated objects that mimicked natural forms but with features and workings beyond anyone's ken.

• • •

Around the time that Homer was describing Hephaestus's intelligent Golden Maidens on Mount Olympus, the poet Hesiod was using similar language to describe their cousin, Pandora. She too was "made, not born." But this female replica was sent down to earth, on a mission from a god.

CHAPTER 8

PANDORA

BEAUTIFUL, ARTIFICIAL, EVIL

TO PUNISH MORTALS for accepting fire stolen from the gods, Zeus commanded Hephaestus to make a "snare" (*dolos*) in the form of a desirable young woman called Pandora. This archaic myth was first written down in two separate poems penned in the eighth or seventh century BC, the *Theogony* and *Works and Days* attributed to Hesiod of Boeotia. It is no surprise to find that humanity's defender Prometheus and his thoughtless brother are both involved in this myth of Zeus's retribution via *biotechne*.

When we last encountered the two Titans, they were making the original humans and animals and doling out natural capabilities, as requested by Zeus (chapter 4). In this mythic cycle of Zeus's revenge on humankind, Prometheus has been freed at last by the hero Heracles from the rock where he was chained. Prometheus and Epimetheus are now the allies and associates of earthlings. Armed with foresight (and rational paranoia), Prometheus tells his impulsive brother to reject any gifts from Zeus. True to his name, "Mr. Afterthought" forgets the warning.[1]

To recap the basic story: Zeus, fuming over the theft of fire, contrives a way to deliver an eternal curse disguised as a gift for humans—a *kalon kakon*, "beautiful evil"—with the help of the smith god Hephaestus. Hephaestus creates an artificial female, a simulacrum or effigy of a woman. Athena and the other gods contribute to her composition, hence her name Pandora, "All Gifts" (the name can mean either "giver" or "recipient"). Dispatched to earth with more nefarious "gifts," a swarm of evil spirits sealed inside a jar, Pandora is the source of all the misfortunes and sorrows suffered by mortals.[2]

As in the Old Testament story of Eve and the serpent, the Pandora myth blames a woman as the agent of mankind's woes. The similarity has elicited much religious and moral soul-searching about patriarchy and the relationship of the sexes in both ancient and modern cultures. Both stories pose profound philosophical questions about theodicy, the existence of evil, divine omniscience and entrapment and human autonomy, temptation, and free will.[3] Yet there are significant differences in the traditions. In the Genesis tale, Eve was an afterthought, created to be a helpmeet for the lonely first man, Adam. The Creator willed Eve to life from Adam's rib and forbade the couple to eat a certain fruit, thus setting in motion a chain of events leading to mortals' original sin. In the Greek myth recounted by Hesiod and others, Pandora is a beguiling artifice deliberately designed by Zeus with gleeful malice toward the human race.

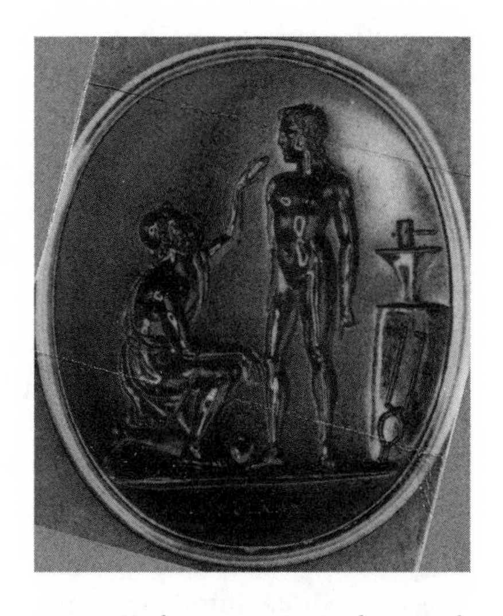

FIG. 8.1. Hephaestus creating Pandora, a modern neoclassical gem commissioned by Prince Stanislas Poniatowski (1754–1833) to interpret the Pandora myth as described by Hesiod. Beazley Collection, photo courtesy of Claudia Wagner.

A crucial difference between Eve and Pandora is that Pandora was not summoned into existence but constructed, by the god of craftsmanship—the same god, Hephaestus, who built other ingenious automata, such as the bronze robot Talos, the self-moving tripods, and a staff of female helpers made of gold (chapter 7). Indeed, Pandora's "manufactured" nature is prominent in all versions of the Greek story, as many classical commentators have pointed out. Pandora's fabrication and her artificiality are also the focus of ancient artistic representations.[4]

• • •

In the brief version in Hesiod's *Theogony* (507–616), Hephaestus, following Zeus's orders, molds the image of a nubile girl. He places on her head a splendiferous crown of gold decorated with *daedala*, intricately worked miniature sea and land monsters so lifelike they seem to writhe and roar. The special crown is reminiscent of the Daedalic sound and light display that Hephaestus crafted on Achilles's marvelous shield, and the vivid artistic images that terrified Odysseus in the Underworld (chapters 7 and 5).[5] Next, Athena dresses this unnamed maiden in a shimmering robe and veil and tucks spring flowers in her hair. Zeus's plot depends on the artificial girl's ethereal physical beauty and her luxurious adornments to "trick" mortals. When Zeus displays the completed Pandora to a gathering of gods and men, everyone is filled with awe (*thauma*). Their reaction—"seized with amazement"—parallels other ancient descriptions of the uncanny emotions evoked by encounters with miraculously realistic statues (chapter 5).[6]

The "manufactured maiden, gift of Zeus," is accepted by "foolish" Epimetheus, who eagerly welcomes her to his home. There is no mention of the jar filled with disasters, and Pandora is not named or called the first woman in the *Theogony*. Hesiod piles on heavy-handed misogyny. Pandora is presented as the prototype of idle, greedy women parasitic on men's labor and economic wealth, like queen bees sponging up nectar stored up by worker bees. Hesiod ends with a jeremiad on "the deadly race of females who live with mortal men" and bring them never-ending misery.

A different tone suffuses the longer, more dramatic episode in Hesiod's *Works and Days* (53–105). Again, Zeus is portrayed as a vindictive tyrant taking malicious pleasure in his plot to make humankind pay forever for the secret of fire. He laughs out loud as he orders Hephaestus to create an android in the form of a seductive virgin that will bring ruin to men even as her charms arouse lust and love. Hephaestus molds clay into the shape of a young woman with the unearthly splendor of an immortal goddess. Like Pygmalion's ivory virgin, "the manufactured Pandora" surpasses the beauty of any mortal woman ever born. Hesiod's descriptions make it clear that Pandora is not a real woman but a "constructed thing."[7]

Zeus instructs Hephaestus to give this bewitching female facsimile the power to move on its own, as well as humanlike strength and voice. Next, the Olympian divinities come forward to bestow unique gifts, capabilities, and personality traits, as commanded by Zeus. Athena teaches Pandora crafts and dresses her in dazzling clothing; the Graces

FIG. 8.2. Hermes presents Pandora to Epimetheus, a cast of a modern neoclassical gem commissioned by Prince Stanislas Poniatowski (1754–1833) to interpret the Pandora myth as described by Hesiod. Beazley Collection, photo courtesy of Claudia Wagner.

and Peitho give her charm and the power of persuasion, while Aphrodite fills her with irresistible sex appeal (Pandora arouses *pothos*, "painful desire and yearning"). Hermes, the trickster-messenger god of thieves and transgressions, gives Pandora a shameless, devious nature and deceitful words. It is Hermes who names her "Pandora, for all the gifts the gods had given her for the ruination of mankind."[8] The "trap is now complete," writes Hesiod, and "the Father of Men and Gods sends Hermes to present the gift to Epimetheus."

Epimetheus assumes Pandora is a real woman. Pandora calls to mind another myth about a cunning artifice that was a dangerous gift—the Trojan Horse. Some versions of the story of the Trojan Horse, built by the Greeks and presented to the Trojans as a ruse of war, suggest that it was sometimes imagined as an animated statue with articulated joints and eyes that moved realistically. It is striking that some tales also recounted ways to determine whether the magnificent horse was real or an artifice. The tests involved piercing its hide to see if it would bleed. But there was no clever riddle or mythic version of the Turing test to help mortals recognize "Artificial Intelligence" in antiquity.[9] Heedless of his brother's warning, writes Hesiod, Epimetheus "took the gift and understood too late."

As a being that was *made, not born*, Pandora is unnatural. A replicant with no past, Pandora is unaware of her origins and her purpose on earth. As a "marvelously animated statue" she exists outside the "natural cycles" of birth, "maturation, and decay." Even the gods, although ageless and undying, were born; they possess memory and have offspring. Like the perfect maiden Galatea molded by Pygmalion and the instantly adult replicants of the *Blade Runner* films, Pandora has no parents, no childhood, no history, no memories, no emotional depth, and no self-identity or soul. Though sometimes thought of as the "first woman," Pandora does not reproduce, age, or die.[10]

In terms of traditional creation beliefs, of course, "all mortals are Pandoras, that is, products of divine artifice."[11] But in the Greek mythic imagination, Pandora was visualized as different from a biological woman; she was a replica of a woman, "a lovely maiden-shape" of clay, made with the same substance and process that craftsmen used to make statues and other objects. Impersonating an adorable, accomplished girl of marriageable age, Pandora is endowed with a low sort of intelligence (Hermes gives her the "mind of a female dog" according to Hesiod, *Works and Days* 67). It is unclear whether Pandora has the ability to learn, choose, or act autonomously. Her only mission is to open the jar of all human misfortune.

An outstanding feature of Hesiod's poems is the similarity between Pandora's creation by Hephaestus and Homer's description of the self-moving, thinking, and talking female androids devised by Hephaestus in the *Iliad*, written around the same time as *Works and Days*. Inner workings or mechanics are not described in either case. But it is striking that Hesiod's language makes Pandora "essentially indistinguishable" from the golden automata described by Homer. Pandora "begins as inert matter—in this case not gold but clay"—and she becomes a "humanoid machine" endowed with mind, speech, and strength, knowledge of crafts from the gods, and the ability to initiate action.[12]

• • •

Ancient artistic illustrations of the Pandora myth center on her fabrication by Hephaestus and her attributes given by the gods. One example is a Campanian amphora, attributed to the Owl Pillar Group, a circle of Etruscan artists who made clumsy but charming copies of Attic vases in the fifth century BC. On one side of the vase Zeus stands looking at Pandora's jar (fig. 8.11) while on the other side, Hephaestus leans on his hammer next to the half-complete Pandora.[13]

The Athenian vase in figure 8.3 (plate 12; about 450 BC) shows a bearded man labeled "Epimetheus" gazing in wonder at Pandora, who flirtatiously tosses her head back and holds up her arms. She is dressed in a bride's finery, but her demeanor is not that of a modest maiden. Their eyes meet and a small winged Eros (sexual desire) flies toward Epimetheus, reinforcing the sexual electricity between them. Behind them, two other figures lock eyes. Hermes—who gave Pandora all of her wicked qualities—turns to look back at Zeus. The two gods seem to be on the verge of smiling as they exchange a conspiratorial glance, reminding the viewer of the chain of trickery about to be played on the hapless Epimetheus and all humankind.[14]

A detail on this vase is puzzling: why does Epimetheus carry a hammer, the signature attribute of Hephaestus? Another vase, attributed to

FIG. 8.3 (PLATE 12). Epimetheus and Pandora, right; on left, Zeus and Hermes exchange a conspiratorial smile. AN1896–1908 G.275 attributed to the Group of Polygnotos, Attic red-figure pottery volute-krater, about 475–425 BC. Image © Ashmolean Museum, University of Oxford.

the Polygnotus Group, shows the upper half of a female, apparently Pandora, flanked by satyrs with hammers. A similar scene appears on a fifth-century BC vase by the Penthesilea Painter, showing dancing satyr and Pan figures around the upper body of a maiden thought to be Pandora. A frieze of dancing satyrs also decorates the majestic vase illustrating Pandora's myth by the Niobid Painter, discussed below. Why the satyrs? Scholars suggest that these images might illustrate a lost satyr play titled *Pandora* or *The Hammerers* by Sophocles. Known only from fragments, the Athenian comedy featured a workshop scene in which a chorus of hammer-wielding satyrs assist Hephaestus in the making of Pandora.[15]

Another notable aspect of the two vases described above is that Pandora's body seems to be emerging from the ground. But Pandora is not a goddess of the Underworld or a chthonic (earthborn) figure. Instead, as some scholars conclude, the image of the upper half of Pandora is intended to indicate that she was molded from earth by Hephaestus's craft.[16] This interpretation could be supported by similar imagery on the Etruscan gems in chapter 6, in which Prometheus is in the process of forming the first human from clay. The gem artists depict the first human as an upper body with a raised arm.

• • •

Other vase painters emphasize the rigid statue-like or doll-like appearance of Pandora, attended by active gods and goddesses. In these images, Pandora is in the process of being made and imbued with human attributes, but she is not yet animated or set in motion. A black-figure amphora attributed to the Diosphos Painter (525–475 BC) appears to be the most ancient representation of Pandora. This interpretation was proposed by Theodor Panofka in 1832, upon the first publication of the vase.

In figure 8.4 (plate 13), we see Zeus, standing with a small doll-like woman in his hands. He appears to be admiring Hephaestus's handiwork, while a goddess holds out wreaths to adorn her and Hermes steps to the right. The Diosphos Painter is known for his unusual iconography and the two inscriptions are nonsense words, which complicates the identification of the figures. Adolf Furtwangler proposed in 1885 that the small stiff figure could be Athena, who was born fully armed with helmet, spear, and shield from Zeus's head. But unlike other vase paintings of the birth

FIG. 8.4 (PLATE 13). Zeus holding Pandora, with goddess (Athena?) and Hermes. Attic black-figure amphora, Diosphos Painter, about 525–475 BC, F 1837. Bpk Bildagentur / Photo by Johannes Laurentius / Antikensammlung, Staatliche Museen, Berlin / Art Resource, NY.

of Athena, this scene includes no helmet or weapons. The goddess presenting wreaths to the figurine appears to be Athena adorning Pandora, as in other vase paintings (see figs. 8.5 and 8.6). The presence of Hermes, Pandora's escort, is also significant. It seems likely that the vase depicts Pandora, as suggested by Panofka.[17]

Pandora's completion is clearly represented inside a large shallow bowl (about twelve inches across) by the Tarquinia Painter (470–465 BC, fig. 8.5), probably made for display as a temple dedication to Athena. Pandora's inscription, *Anesidora*, gives her alternative name, "She who releases gifts." Unfortunately, the black, brown, and purple painting on white ground is damaged, but one can see how Pandora stands passively like "an inanimate, created object" between the taller active gods, Athena and Hephaestus, who are putting the finishing touches on their creation.[18] Posed as a "lifeless" mannequin with feet together and hands "hanging limply at her sides," Pandora head is turned toward Athena.[19] Athena is

FIG. 8.5. Hephaestus (right) and Athena (left) placing finishing touches on Pandora (center), red-figure Attic cup from Nola, about 470–460 BC, Tarquinia Painter, inv. 1881,0528.1. © The Trustees of the British Museum.

fastening the shoulder of Pandora's gown and Hephaestus is placing a crown on her head (his trusty hammer is in his left hand). The scene replicates the way statues were offered gifts, dressed in finery, and adorned with jewelry in antiquity.[20]

• • •

The image of Pandora is even more striking on a superb, oversized krater, more than a foot tall, by the Niobid Painter (about 460 BC, figs. 8.6 and 8.7, plate 14). Pandora's stiff posture and facial expression reinforce her artificial status and her fatal attraction. She stands within a V created by spears, and the V shape is repeated in the decorative top border of the vase. That border has a rare motif pattern that resembles a set of craftsman's tools, tongs like those used by Hephaestus and blacksmiths in other vase paintings (see figs 7.4 and 7.5). This uniquely appropriate detail reinforces the idea that Pandora was *made, not born.* The same tool motif also appears prominently in the border around the top of the great vase of about 440 BC that depicts the death of the bronze robot Talos—who was also crafted by Hephaestus (see fig 1.3).[21]

In the Niobid Painter's vase scene, Pandora stands like a motionless wooden *xoanon* idol or a marble statue with her arms at her sides, looking straight ahead. The vase scholar H. A. Shapiro likens her to a "wind-up doll" waiting to be wound up. There is a flurry of activity around Pandora. Athena approaches from one side holding out a wreath, with Poseidon, Zeus, and Iris lined up behind her. On Pandora's other side we see Ares, Hermes, and Hera (or Aphrodite). The lineup includes some gods not

FIG. 8.6. Pandora admired by gods and goddesses, on the magnificent red-figure calyx krater, by the Niobid Painter, about 460 BC, inv. 1856,1213.1. © The Trustees of the British Museum.

mentioned by Hesiod as contributors to Pandora's manufacture. Moreover, the gods appear to be talking among themselves and reacting to Pandora, instead of presenting endowments. The scene probably illustrates the later passage in Hesiod, "when Zeus shows off his new plaything to the Olympian gods before inflicting her on mankind."[22]

Pandora stares straight ahead. In conventional vase painting iconography, the faces of gods, people, and animals are almost always shown in profile or three-quarter views; views of human faces from the front are very rare. In Greek art, a full-frontal face indicates a kind of mindlessness, used for dead or nonliving figures and especially for masks and statues. Frontal views can also suggest a mesmerizing gaze. Notably, the Niobid Painter, known for his elegantly simple classical style, employs frontal faces for dead and dying figures in two of his other famous vases, the Geta Krater, showing Greeks killing Amazons, and his name vase, showing the massacre of Niobe's children.[23] In the arresting frieze illustrating the Pandora myth, both effects—a blank mind and a compelling stare—seem to be intended by Pandora's forward-facing stance.

The scene holds yet another remarkable element. Facial expressions showing emotion, such as grimaces, frowns, or smiles, are also very rare in Greek vase paintings. People's faces in vase paintings are usually impassive, with emotions indicated by gestures or posture.[24] But this exceptional Pandora not only faces forward, gazing out at the beholder; she is smiling. What message does her smile send? A broad smile strikes one as inappropriate for a virginal bride—but recall that Hesiod described Pandora as a shameless and seductive animated statue. Pandora's unexpected expression could remind ancient observers of the face of a *kore*, a life-size painted marble statue of a young, draped maiden typical of the archaic period (600–480 BC). The lips of a *kore* statue (and those of her counterpart, a nude male *kouros*) invariably curve up in a curiously mirthless smile.

The same incongruous smile appears on the implacable faces of archaic marble statues depicted in scenes of violence.[25] The preternaturally serene—some would say vacuous—expression on archaic statues is known by art historians as "the archaic smile." With her statue-like stance and that faintly creepy smile, the Niobid Painter underscores Pandora's manufactured origin and portrays her as an automaton at the moment of her animation.

FIG. 8.7 (PLATE 14). Detail, Pandora admired by gods and goddesses, on the red-figure calyx krater by the Niobid Painter, about 460 BC, inv. 1856,1213.1. © The Trustees of the British Museum.

FIG. 8.8. Kore statue with enigmatic "archaic smiles." Left, the Peplos Kore, painted marble, about 530 BC, Acropolis Museum, Athens, HIP / Art Resource, NY. Top right, head of the Peplos Kore, photo by Xuan Che, 2011. Bottom right, marble Kore head, sixth century BC, Musées Royaux d'Art et d'Histoire, Brussels. Werner Forman / Art Resource, NY.

The scene on this magnificent vase—with the unusual "special effect" of the artificial young woman staring fixedly out at the viewer wearing a disconcerting smile—must have had a strong impact on viewers more than twenty-four hundred years ago. The smiling automata would intensify an Uncanny Valley response.

This image of a leering Pandora resonates with a modern cinematic sister of Pandora, the evil, smirking automaton Maria in the brilliant silent film *Metropolis* of 1927. Widely recognized as one of the most influential science-fiction films in cinema history, the director Fritz Lang's tour de force features grim expressionist cityscapes and special-effects technology staggering for the 1920s and still stunning today. *Metropolis* envisions a future dystopia ruled by the rich, who dominate the impoverished masses with demonic machines.[26] The publicity photos showing the robot Maria with her makers and the actress being prepared for her scene have startling similarities to the ancient vases depicting Pandora being groomed by the gods before her big scene on earth.

Filmed only seven years after the word *robot* entered the popular lexicon, *Metropolis* stars an erotic femme fatale robot deliberately created to wreak havoc in the world. The film, made as the pace of machine technology and industrialization was escalating in Europe and America, shows how swiftly the novel ideas of robots and the merging of humans and machines captured the popular imagination. Critics note that the film's story line is riddled with illogical twists. But so is the ancient myth of Pandora. Yet, as with the other ancient tales of artificial life gathered in this book, the message is clear. With each new generation, the age-old opposition of human versus machine continues to exert an edgy push-pull response, trepidation commingled with fascination and awe.

In the Greek myth, Pandora's deceptive appearance as a "tender maiden" is designed to delight and seduce men while bringing them endless suffering. In *Metropolis* a sweet young woman (Maria, played by a seventeen-year-old actress) is transformed into a sexualized robot-vamp designed to bring chaos and disaster. In a spectacularly filmed sequence of futuristic technology involving crypto-chemistry and pulsating rings of "electrical fluid," the robot's metallic form is animated by draining the life force of the innocent young woman encased inside. The "electrical fluid" recalls the ichor of Talos (chapter 1) and the electricity that animates Frankenstein's monster (chapter 6).[27]

FIG. 8.9. The evil *Maschinenmensch* (machine-human) Maria with her makers, in Fritz Lang's *Metropolis* (1927). Production still courtesy of metropolis1927.com. Scene from *Metropolis* film, Adoc-photos / Art Resource, NY.

In the film, Maria's diabolical robotic doppelgänger is characterized by her hypnotic, "slow, irresistible movements" and an inhuman "basilisk motion of the head." Like the strangely grinning automaton Pandora on the vase by the Niobid Painter, the artificial Maria's "haunting loveliness" is accompanied by a "weird, incomprehensible smile."[28]

• • •

Other paintings by the innovative Niobid Painter are believed to have been influenced by wall paintings in classical Athens. Was his scene of Pandora also based on a painting of similar composition in the city? That is unknown. But we do know that Pandora's creation by Hephaestus was of such importance in Athens that it was displayed in a key location on the Acropolis. A similar "lineup" of gods and goddesses on either side of Pandora appeared in relief on the massive pedestal of the colossal gold and ivory statue of Athena inside the Parthenon.[29] This masterpiece was the work of the famed sculptor Phidias in 447–430 BC. According

FIG. 8.10. Interesting coincidences in the ancient and modern portrayals of an evil female robot. Top left, Pandora as a stiff automaton being prepared by the gods for her mission on earth (Niobid Vase, fifth century BC) and the actress being groomed for her role as the robot Maria in the film *Metropolis* (1927). Right, Pandora and Maria robot. Bottom, the transformation of Maria into a robotic winking and smirking double. Last image, Hope/Elpis with crooked smile, sixth century BC. Photo collage by Michele Angel.

to Pliny (36.4), writing in the first century AD, the scene on the base depicted Pandora attended by twenty gods and goddesses, who would have been nearly life-size.

A century later Pausanias (1.24.5–7) also admired the imposing statue of Athena and the scene of Pandora's creation on the Acropolis. The original colossus and base are lost, but one can begin to visualize them based on a large marble copy of the base made in about 200 BC, found in 1880 in the ruins of Pergamon (Turkey). A small marble Roman replica of the statue and the base (first century AD) also came to light on the Athenian Acropolis in 1859. These artifacts make it "clear that Pandora was shown as a statue-like figure," created and adorned by Hephaestus and Athena, who were venerated together in Athens as the patrons of arts and craft.[30]

Further evidence of the scene's popularity in Athens was discovered in the Athenian Agora. Since 1986, fragments have been excavated there of another public image of the creation of Pandora attended by divinities on a marble relief. Among the figures found so far are Hephaestus and Zeus. The archaeologists have also unearthed the marble head of a woman. Who is she? One clue is her oddly disconcerting smile—but her identity, revealed below, is surprising.[31]

• • •

In the myth, Pandora was escorted to earth by Hermes and presented to Epimetheus as his bride. Zeus knew that Prometheus's brother lacked foresight and good judgment, making him the perfect patsy. Pandora's "dowry" was a sealed *pithos*, a large jar used for storage. Hesiod calls the *pithos* "unbreakable," an adjective usually applied to metal, so the jar was probably originally imagined as bronze. It seems that *pithos* was mistranslated as *pyxis* (box) in the sixteenth century, and since then the image of Pandora's box persists in the popular imagination. No ancient artworks show Pandora with the jar of troubles or actually opening the *pithos* and reeling back in horror, but those scenes are favored in more than a hundred medieval and modern retellings in poems, novels, operas, ballets, drawings, sculptures, paintings, and other artworks. The series of neoclassical sculpted reliefs and drawings by John Flaxman (1775–1826) illustrating vignettes from Hesiod's Pandora were immensely popular at the end of the eighteenth century, when the antiquarian carved gems in figs 8.1 and 8.2 were also created.[32]

The contents of the forbidden *pithos*, all the misfortunes that afflict the mortal world, were unknown to Pandora. But Zeus was counting on her to open the jar, releasing disease, pestilence, endless labor, poverty, grief, old age, and other dire torments on humanity forever.[33] Pandora's jar of evils seems to be related to the passage in Homer's *Iliad* (24.527–28) describing two fateful jars kept by Zeus. One urn is filled with blessings, the other with misfortune, and the contents were randomly mingled and showered upon humans by Zeus. Presumably, it is Zeus's *pithos* of misery and evil that accompanies Pandora. She "serves as his agent for opening the jar."[34]

In the myth recounted by Hesiod (*Works and Days* 90–99), once in Epimetheus's house, Pandora lifts the lid of the great *pithos*, and

FIG. 8.11. Zeus contemplates Hope/Elpis peeping out of Pandora's jar. Red-figure amphora from Basilicata, fifth century BC, inv. 1865,0103. © The Trustees of the British Museum.

the evils swarm out. When the lid is slammed down—by Pandora's hand but by Zeus's design—one spirit is trapped inside. This is *Elpis*, "Hope." The meaning of this crucial detail has been intensely debated since antiquity.

In antiquity, Elpis/Hope was personified as a young woman. In "The Girl in the *Pithos*" (2005), classical archaeologist Jenifer Neils identifies three ancient artifacts that represent Elpis in Pandora's jar. The first was, until 2005, the only known image of Elpis. It appears on the Owl Pillar Etruscan amphora mentioned above, with one side depicting Hephaestus and the half-completed Pandora, the beginning of the myth. The other side of that vase illustrates how the story ends (fig. 8.11).

FIG. 8.12. Grinning Hope/Elpis peeking out of Pandora's jar. Aryballos (perfume flask), ceramic, sixth century BC, Thebes, Boeotia, Greece. Henry Lillie Pierce Fund, 01.8056. Photograph © 2018 Museum of Fine Arts, Boston.

A bearded Zeus contemplates a large *pithos* with a small girl peeping out of the jar. She is Elpis/Hope, confined in the *pithos* by Zeus's order. This intriguing vase surely copied a more sophisticated Attic vase now lost, notes Neils. The Etruscan artist "juxtaposes two analogous scenes." In each vignette, a male divinity "contemplates female evil."[35]

The second artifact is a small terra-cotta *aryballos* (perfume flask) from Boeotia, a region north of Athens, made in about 625–600 BC. It is shaped like a *pithos* with the sculpted head of a young woman at the top as though popping up out of the jar (fig. 8.12). The opening of the flask is made to look like the lid of the jar. We can assume, with Neils, that the potter was inspired by his fellow Boeotian Hesiod's description of Elpis/Hope in *Works and Days*, written some years earlier, in about 700 BC. The *aryballos* held perfume, remarks Neils, a substance, like Pandora's charms, that was considered a seductive snare for men, suggesting a humorous or ironic spin on the myth.[36]

There is plenty of evidence that the sophisticated ancient Greeks appreciated both tragedy and comedy in Pandora's story. Sophocles's lost satyr play and the vases juxtaposing satyrs with Pandora are some examples of a lighthearted approach. Hesiod says Zeus laughed while devising his trick on man, and amusement is implied on the vase showing Zeus and Hermes enjoying the joke on Epimetheus (fig. 8.3, plate 12). The Niobid Painter's vase continues the sardonic theme with a broadly smiling Pandora (fig. 8.7, plate 14). Take a closer look at the young woman popping out of the little perfume jar in fig. 8.12. She wears an ironic lopsided grin, a sly smirk.[37]

The third likely image of Elpis/Hope was found among the fragments of the fifth-century BC high-relief panel discovered in the Athenian Agora, mentioned earlier. Archaeologist Evelyn Harrison identified the frieze as an illustration of the Pandora myth. Along with the marble figures of Hephaestus and Zeus, archaeologists found a female head with a "strange, slightly wicked expression," an asymmetrical smile. But, to answer the question posed above, she is not Pandora—the disembodied head is larger than the heads of the figures of the gods and it is flat on top. Neils proposes that this head belonged to a figure of Elpis/Hope peeping out of a large *pithos*. "Facial expressions are extremely rare in Greek art," comments Neils, "but a smirk seems a particularly apt way to characterize the personification of false hope."[38]

• • •

Was Elpis/Hope a blessing or an evil? The mythic traditions about Pandora are labyrinthine; several aspects of the story as it survives in ancient literature and art strain logic.[39] In particular, the vexing question of why Hope remained in the jar has bedeviled commentators ever since the myth was first told. The enigmatic smiles of Pandora and Elpis seem to mock attempts to untangle the puzzle.

Hesiod is ambiguous: Is Hope one of the troubles in the pack of evils dispersed in the world? Or is Hope humans' only solace now that their world is so troubled? The modern fairy-tale version of the myth casts Hope as a merciful spirit that remained behind to comfort humans or a blessing bestowed by Zeus to compensate for the evils. But keep in mind that the ancient Greeks generally considered Hope to be negative or misleading, as is evident in the common epithet "blind hope." Notably, Hesiod (*Works and Days* 498, 500) calls Elpis/Hope "empty" and "bad." In the *Iliad* (2.227) Athena plants false hope in the mind of the doomed Trojan hero Hector before he is killed in the duel with Achilles. The fifth-century BC poet Pindar (frag. 214) says Elpis/Hope "rules man's ever-changeable mind." Aristotle is not much help: he defines *elpis* as the "future-directed counterpart of memory," connoting the ability to anticipate good or evil consequences.[40]

In the fifth-century BC Athenian tragedy *Prometheus Bound* (128–284), Prometheus confesses that he gave mortals another gift along with fire: he deprived them of the ability to "foresee their doom (*moros*)" by "causing blind hopes (*elpides*) to live in their hearts," so that they will persevere. The play only intensifies the philosophical questions surrounding the existential meaning of hope. It seems that in the new, harsher world of the present, humans have come to resemble Prometheus's brother Epimetheus, lacking the ability to see what lies ahead. Is such an illusion a boon or a curse?[41]

The ambiguity of Hope's meaning in antiquity compounds the enigma of Pandora's *pithos*. In the murk of the myth as it has come down to us, we can set out the following seemingly contradictory options: The contents of the jar are evil, and they are activated by being released to bring harm to humans. Hope is not let out: either it is an evil that harms humans like the other things in the jar, or it is unlike the evil things in the jar and is good for us. So hope is either activated, like the other evils, despite being kept in the jar, or hope is not activated because confined inside the jar.

Four possible scenarios can be posed: (1) Hope is *good*, despite being in the jar of evils, and *activated* by Zeus to offset evils; (2) Hope is *good* but is trapped inside the jar by Zeus, therefore further harming humans; (3) Hope is one of the *evils* in the jar and *activated*, despite being trapped in the jar, and is meant to torment humans with wishful thinking and illusion; (4) Hope is *evil* but *not activated*; it is trapped by Zeus in order to spare humans from false hopes.[42]

The mystery of Elpis/Hope trapped in the jar of evils resists resolution. The best interpretation may be that Hope is neither all good or all bad, nor is she neutral. Hope is a uniquely human emotion. Like the artificial woman Pandora, Elpis/Hope represents a *kalon kakon,* beautiful evil, a seductive snare, beckoning irresistibly while hiding inherent and potential disasters.

This dilemma was devised more than two millennia ago in the context of artificial life created by an ingenious inventor with surpassingly superior biotechnology; its ambiguity could not be more pointed for our own era.[43] Who can resist opening Pandora's box of tantalizing "gifts," marvelous science and technology that promise to improve human life? Like Epimetheus, oblivious to the moral and social dangers lurking within, ignoring the warnings of the lone Promethean voices among us, we rush headlong into a future of humanoid robots, brain-computer interfaces, magnified powers, unnaturally enhanced life, animated thinking things, virtual reality, and Artificial Intelligence. We blunder on, hoping for the best.

● ● ●

Two millennia before Isaac Asimov conceived of the Laws of Robotics (1942), the ancient Greek mythologists imagined animated statues set in motion and imprinted with specific missions to help or harm. Asimov's original three laws specified that (1) *a robot may not injure a human being*; (2) *a robot must obey orders given by humans unless this would cause harm to a human*; and (3) *a robot must protect itself unless this conflicts with laws 1 and 2*. As we've seen, Hephaestus surrounded himself with benign automata and self-moving tripods to make his life easier, and he gave the world happy marvels such as the singing maiden statues at Delphi. But Hephaestus was capable of manufacturing harmful artifices too, beginning mildly with the throne that trapped his mother, Hera, and culminating in Pandora, his crowning and awful achievement commissioned by the all-powerful Zeus. In myth, Talos the bronze robot, the dragon-teeth army, the mechanical eagle, the fire-breathing bulls—all were deliberately intended to injure humans, breaking Asimov's first law.[44]

Pandora certainly flouts rule number 1. But the scale of her devastation is so vast—the ruination of all humankind, as plotted by the tyrant

Zeus—that Asimov's fourth law applies. Pandora breaks the so-called Zeroth Law, which Asimov added later: *a robot shall not harm humanity.* Pandora also violates law 23 of the 2017 Asilomar principles: *Artificial Intelligence should benefit all humanity* (chapter 7).

One cannot help noticing that all of the automata used to inflict pain and death in ancient mythology belonged to tyrannical rulers, from King Minos of Crete and King Aeetes of Colchis to Zeus, the Father of Gods and Men, who chuckles in anticipation of his cruel "trap" for humans. It is a striking fact that the autocratic fascination with animated statues designed to inflict torture and death was not confined to ancient myth. Malevolent machines existed in reality—in historical times—and were used by living tyrants of the ancient world. The next chapter surveys actual automata and self-moving devices—some designed to harm and others created for benign purposes—described in literature, history, legend, and art beginning as early as the fifth century BC.

CHAPTER 9

BETWEEN MYTH AND HISTORY

REAL AUTOMATA AND LIFELIKE ARTIFICES IN THE ANCIENT WORLD

SO FAR WE have considered how the ancient Greeks imagined—through mythology and artworks—artificially created life, animated statues, beings that were not biologically born but manufactured, fantastic technologies, and augmented human powers. We saw how people in antiquity portrayed Daedalus, Medea, Prometheus, and Hephaestus as super-geniuses, picturing them employing familiar tools and methods but with miraculous capabilities to construct marvelous things far beyond what could be achieved by mortals.

Except for the bronze robot Talos and the first humans made by Prometheus, practical details and inner workings of divinely crafted artifices are missing in the mythic narratives and fragments that have come down to us. But the wide of range of stories about *biotechne* reveal that the idea of *making* artificial life was conceivable in antiquity, portrayed as stupendous feats of ingenuity and craft. Some divine devices in myth might have arisen as metaphors for innovations in technology, while others may have been exaggerations of more modest counterparts in historical times. Earthly, simple approximations of some of the mythical marvels might have been practicable with available tools, materials, techniques—and formidable intelligence—in antiquity. Even so, it is important to resist the temptation to project modern motivations and assumptions about technology onto the ancient world.[1] Although many of the ancient myths and ideas about artificial life certainly call to mind and seem to foreshadow later inventions, one cannot project direct lines of influence from antiquity to modern biomechanics and robots.

The history of real mechanical designs and practical inventions, from artillery, the catapult, and theatrical technologies involving pulleys, levers, springs, and winches to self-operating devices, from the Mediterranean world to China, has been intensely and comprehensively studied.[2] From the wealth of well-documented ancient concepts and designs of automata and machines in the history of ancient technology, I have selected examples for this chapter that echo or resound in some way with the self-moving objects, animated statues, and other ways of imitating life from the realms of mythology discussed in the previous chapters. As we move from myth to history, keep in mind that it is inevitable that elements of popular folklore and legend have seeped into some surviving and fragmentary accounts of actual inventions. The historical incidents in the following pages do not constitute an exhaustive survey but are meant to give an idea of the various kinds of lifelike replicas and automata—some deadly, some grandiose, others charming curiosities—that were really designed and/or tested between the sixth century BC and about AD 1000.

Historians of robotics suggest that automata fall into three basic functions: labor, sex, and entertainment or spectacle. These features appeared in the ancient myths and legends about artificial life. Self-operating devices resembling living beings could be used to amplify human capabilities, to dazzle and awe, to trick and deceive, to injure and kill. Automata could serve as trappings and manifestations of power, sometimes in benign ways but other times with malicious intent.

In Greek myths, Zeus is portrayed as a spiteful tyrant who takes joy in devising a hideous torture for Prometheus and dispatches the seductive artificial woman, Pandora, to inflict suffering on all humankind. These torments required the technological expertise of Hephaestus, who also constructed King Aeetes's bronze bulls, to burn Jason, and King Minos's bronze killer Talos. A pattern stands out in these and other myths about devices made to inflict pain and death: each artifice was commissioned and/or deployed by a despotic ruler, as a means of displaying arbitrary absolute power. As it turns out, a similar pattern can be traced in historical antiquity: a good number of real tyrannical rulers used wickedly clever contraptions and artifices that mimicked nature to humiliate, harm, torture, or even kill their subjects and enemies.[3]

• • •

As Ovid (*Metamorphoses* 8.189) envisioned the myth, Daedalus created his human enhancement of flight by imitating the power of birds. He made rows of real feathers, assorted by size in a curve, and arched the structures to imitate real bird wings. Then, attaching them to his back and arms, he "balanced his body between the wings and hung poised, beating the air." Unlike the supernatural, effortless flight of the gods that defied time, physics, and space, however, his artificial wings required the physical effort of pumping one's arms to soar like a bird.

For a human being to attempt to fly by flapping man-made wings is of course aeronautically unsound, sure to end badly. That brute fact figured in a sadistic punishment using imitation bird wings meted out annually in ancient Leucadia (modern Lefkada), an Ionian island famed for its sheer sea cliffs. There, the ancient Greeks had "one regular opportunity to experiment with such flying devices without keen regard to safety."[4] Strabo (10.2.9) described the ancient custom on Leucadia known as Criminal's Leap. Each year, as a sacrifice to Apollo, the Leucadians would force a condemned man to "fly" from the island's white limestone cliff (the cliff was later known as Sappho's Leap, after the poetess's fabled suicide, and is now called "Lovers' Leap").[5] Like Icarus of myth, the man was fitted with a pair of artificial wings. And for good measure, all sorts of live birds were fastened to him as well, to add to the spectacle. Spectators on the cliff and in small boats below watched the hapless victim flapping with all his might while surrounded by helplessly fluttering birds.

During the Roman Empire, it was a popular sport to demean, torture, or execute people in amusing scenarios that re-created tragic Greek myths. The emperor Nero was a master of such perverse public entertainments in the Circus and at his banquets (AD 54–68). Two such performances were related by the imperial historian Suetonius (*Life of Nero*). For the play called *The Minotaur*, the individual forced to play Pasiphae was made to crouch "inside the hindquarters of a hollow wooden heifer" while an actor disguised as a bull mounted her. For a ballet reenacting the myth of Daedalus and Icarus, Nero commanded the man cast in the role of Icarus to fly with his artificial wings from a high scaffold. Suetonius records that the man fell "beside Nero's couch, splattering the emperor with blood."

Contriving artificial human enhancements based on bird wings for torture and entertainment was not confined to the ancient Mediterranean world. In China, Gao Yang/Wenxuan, the first emperor of the Northern Qi dynasty in AD 550–559, was feared for his erratic bloodthirsty rages. He enjoyed executing prisoners by harnessing them to great wings woven of bamboo or paper kites in the form of birds, large enough to carry a man. He forced the victims to "fly" from the 108-foot-high Tower of the Golden Phoenix (in the Qi capital, Ye) and laughed at the spectacle of doomed men attempting to stay aloft. Apparently the killer kites were also manipulated by skilled men on the ground holding the strings—the idea was to keep the victim in the air as long as possible. It was reported that hundreds of involuntary "test pilots" died for the emperor's amusement. But one man, Yuan Huangtou, an Eastern Wei prince, won fame for surviving the ordeal in AD 559. Strapped to an ornithopter kite shaped like an owl, he managed to take off from the Phoenix Tower and glided a mile and a half to the Purple Way at Zimo, where he landed safely. Presumably he was aided by the kite-holders on the ground.[6]

• • •

In the Greek myth, Daedalus escaped from King Minos of Crete by flying to Sicily with his bird wings. As we saw, once in Sicily Daedalus continued to create wonderful inventions for King Cocalus in Acragas, including the boiling hot pool used to murder Minos (chapter 5). Daedalus also designed an amazing temple and the impregnable citadel at Acragas for his royal patron. With these mythic stories in mind, we turn to a real-life inventor in the actual history of the city of Acragas (Agrigento). This inventor constructed a torture apparatus for the tyrant of Acragas that bears some resemblances to certain mythic creations by Daedalus and Hephaestus.

Acragas was founded by Greeks from Crete and Rhodes in about 580 BC. An ambitious, wealthy citizen named Phalaris undertook the construction of the grand temple to Zeus Atabyrios (named for the highest peak on Rhodes) at Acragas. Phalaris parlayed his status into military power and became an absolute dictator. Detested for his savage brutality, Phalaris was finally overthrown in 554 BC. During his iron rule, a shrewd Athenian bronze smith named Perilaus, seeking favor with Phalaris and knowing his penchant for torture, forged a lifelike statue of bronze bull. It was hollow, with a trapdoor or hatch big enough for a man to enter.

Perilaus presented this handsome bull statue to Phalaris and explained how it worked. "Should you wish to punish someone, lock him inside the bull and build a fire under it. As the bronze bull's body heats up, the man roasts within!" Then Perilaus described the fiendish mechanism in the bull's interior. Perilaus had installed a system of pipes to amplify the victim's screams. While smoke flowed from the bull's nostrils, the tubes directed the sounds of the victim to issue from bull's mouth, transforming the shrieks of agony into the "most pathetic bellowings of a bull, music to your ears." Impressed, Phalaris slyly requested a demonstration of the special sound effects. "Come then, Perilaus, show me how it works." As soon as Perilaus crept inside to yell into the pipes, Phalaris locked the door and built a fire under the bull. The bronze smith was roasted to death (some say he was baked and then thrown from a cliff).

The story evokes the ironic folk motif of an inventor/criminal killed by his own invention/plot. Yet such sadistic behavior in real-life despots is hardly unknown (two Roman examples were the emperors Nero and Caligula). The existence of the Brazen Bull of Phalaris is not in doubt; it was described in numerous extant and lost ancient sources. And Phalaris became the prototypical evil dictator. In fifth-century Greece, the poet Pindar could assume that everyone knew the "hateful reputation" of Phalaris who, "with his pitiless mind, burned his victims in a bronze bull" (*Pythian* 1.95). A century later, Aristotle twice referred to Phalaris's tyrannical rule as common knowledge.[7]

In the first century BC, Plutarch told of Phalaris's bronze bull in which he burned people alive, citing an earlier lost historian. The historian Diodorus Siculus also expounded on the bull. Pliny (first century AD) criticized the sculptor Perilaus (Perillus) for conceiving of such a horrid use for his art and approved of his fate as the bull's first victim. According to Pliny (34.19.88) the sculptor's other statues were still preserved in Rome "for one purpose only, so that we may hate the hands that made them." In the second century AD, the satirist Lucian composed a humorous essay pretending to defend the reputation of the loathsome Phalaris.[8]

The bull spawned other roasting devices. Plutarch's *Moralia* referred to a lost history by Aristides, who described a very similar Sicilian invention in the city of Segesta, but in the shape of a realistic bronze horse, forged by one Arruntius Paterculus for a cruel tyrant named Aemilius Censorinus, known to reward artisans for inventing novel tortures.[9] Diodorus Siculus, a native of Sicily, mentions another deadly statue, this time in the form of a bronze man, also set up in Segesta but by the vicious tyrant Agathocles, who ruled in about 307 BC (Diodorus 20.71.3; see fig. 5.1, plate 6, for the celebrated Bronze Ram of Syracuse, which belonged to Agathocles).

Diodorus returns to the infamous Brazen Bull of Acragas several times in his history. He notes (19.108) that the statue was located on Phalaris's stronghold, a hill on Cape Ecnomus ("wicked, lawless"). Diodorus describes how during the First Punic War, the Carthaginian general Hamilcar Barca looted costly paintings, sculptures, and other artworks from the cities of Sicily. The most valuable prize was the Brazen Bull of Phalaris in Acragas, which Hamilcar shipped to Carthage (Tunisia) in 245 BC. A century later, at the end of the Third Punic War, the Brazen Bull actually returned to Acragas. When the Roman general Scipio Aemilianus finally defeated Carthage in 146 BC, he restored all the plundered treasures to the cities in Sicily, including the Brazen Bull. Polybius (*Histories* 12.25), writing in the second century BC, confirms that the bellowing bronze bull was taken to Carthage and later returned; Polybius notes that the trapdoor on the bull's back was still operative in the second century BC. In 70 BC, Cicero (*Against Verres* 4.33) states that among the treasures recovered by Scipio from Carthage was the great Brazen Bull of Acragas, which "the most cruel of all tyrants, Phalaris, had used to burn men alive." Scipio took that occasion to observe that the bull was a monument to the barbarism of local Sicilian strongmen, and that Sicily would be better off ruled by the more kindly Romans. Diodorus goes on to affirm that one could still view the notorious Brazen Bull in Acragas, when he was writing his history, sometime in 60–30 BC.[10]

The Brazen Bull of Phalaris continued to exert a morbid appeal into the Middle Ages. According to Christian legends, the martyrs Eustace, Antipas, Pricillian, and George were each burned in a variety of red-hot bronze or copper bull statues in the first to fourth century AD. The final incident appears in Visigoth chronicles, and this time the victim was a hated despot. Burdunellus, tyrant of Zaragosa, Spain, was executed in Toulouse in AD 496 by being "placed inside a bronze bull and burnt to death."[11]

• • •

1

PLATE 1 (FIG. 1.4). "Death of Talos," Ruvo vase detail. Album / Art Resource, NY.

PLATE 2 (FIG. 1.5). Medea watches as Jason uses a tool to unseal the bolt in Talos's ankle held by a small winged figure of Death, as Talos collapses into the arms of Castor and Pollux. Red-figure krater, 450–400 BC, found at Montesarchio, Italy. "Cratere raffigurante la morte di Talos," Museo Archeologico del Sannio Caudino, Montesarchio, per gentile concessione del Ministero dei Beni e delle Attività Culturali e del Turismo, fototeca del Polo Museale della Campania.

2

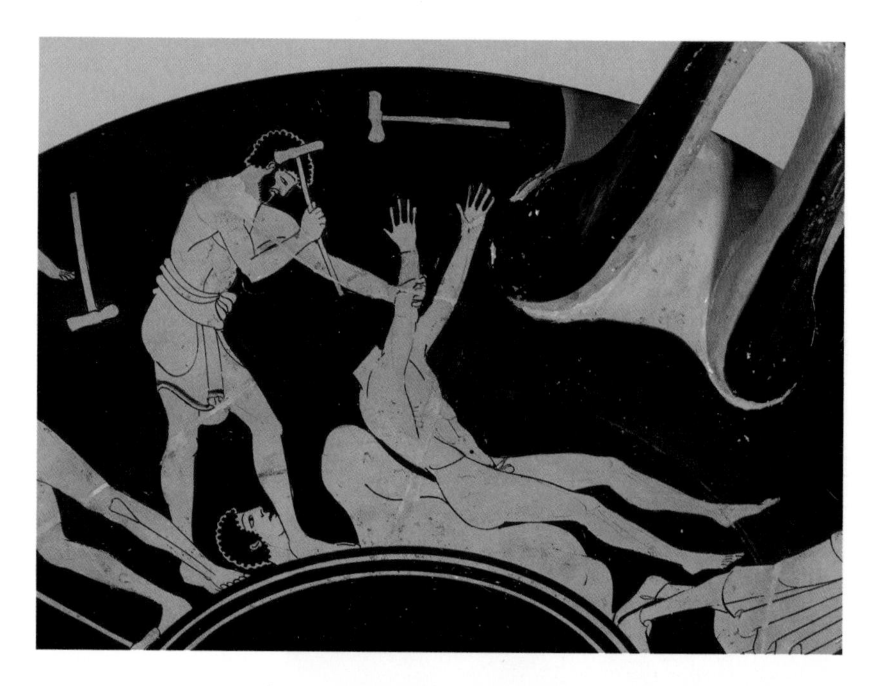

PLATE 3 (FIG. 1.9). Foundry scene, artisans making a realistic bronze statue of an athlete, in pieces, surrounded by blacksmith tools. Attic red-figure kylix, from Vulci, about 490–480 BC, by the Foundry Painter. Bpk Bildagentur / Photo by Johannes Laurentius / Antikensammlung, Staatliche Museen, Berlin / Art Resource, NY.

PLATE 4 (FIG. 7.4). Blacksmith at work, with tools, red-figure kylix, late sixth century BC, 1980.7. Bpk Bildagentur / Photo by Johannes Laurentius / Antikensammlung, Staatiche Museen, Berlin / Art Resource, NY.

PLATE 5 (FIG. 2.1). Medea, looking back at old Pelias (left), waves her hand over the ram in the cauldron. Jason places a log on the fire, and Pelias's daughter, right, gestures in wonder. Attic black-figure hydria, Leagros Group, 510–500 BC, inv. 1843,1103.59. © The Trustees of the British Museum.

PLATE 6 (FIG. 5.1). Realistic bronze ram. Was the sculptor of this life-size ram inspired by the story of Daedalus's true-to-life ram dedicated to Aphrodite in the time of King Cocalus? Bronze Ram of Syracuse, Sicily, third century BC, Museo Archeologico, Palermo, Scala / Art Resource, NY.

PLATE 7 (FIG. 5.5, LOWER RIGHT). Athlete, fourth to second century BC, recovered off the coast of Croatia in 1996, Museum of Apoxyomenos, Mali Losinj, Croatia. Photo by Marie-Lan Nguyen, 2013.

PLATE 8 (FIG. 7.7, TOP). Hephaestus (Sethlans) and assistant (Etule) making an artificial horse (Pecse), Etruscan bronze mirror, fourth century BC, from Orvieto, BnF Cabinet des Medailles, Bronze.1333.

PLATE 9 (FIG. 7.8). Athena making a clay model of a horse; she is holding a handful of clay and there is a pile of clay at her feet. Above left, a saw, drill, and bow drill. The horse's back leg is unfinished. Athenian red-figure wine jug, about 460 BC, F 2415. Bpk Bildagentur / Photo by Johannes Laurentius / Antikensammlung, Staatliche Museen, Berlin / Art Resource, NY.

PLATE 10 (FIG. 6.8). Prometheus, seated, constructing the first human skeleton, using a mallet to attach the arm bone to the shoulder. Carnelian intaglio gem, date unknown, perhaps Townley Collection, inv. 1987,0212.250. © The Trustees of the British Museum.

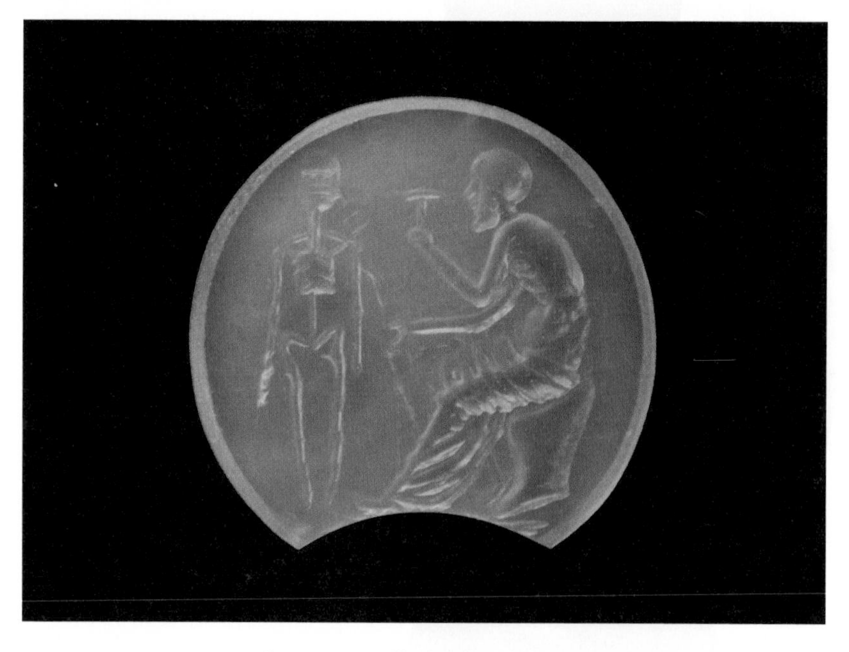

PLATE 11 (FIG. 6.11). Prometheus using a mallet to make a skeleton, chalcedony gem, first century BC, Thorvaldsens Museum, Denmark, acc. no. 185.

PLATE 12 (FIG. 8.3). Epimetheus and Pandora, right; on left, Zeus and Hermes exchange a conspiratorial smile. AN1896–1908 G.275 attributed to the Group of Polygnotos, Attic red-figure pottery volute-krater, about 475–425 BC. Image © Ashmolean Museum, University of Oxford.

PLATE 13 (FIG. 8.4). Zeus holding Pandora, with goddess (Athena?) and Hermes. Attic black-figure amphora, Diosphos Painter, about 525–475 BC, F 1837. Bpk Bildagentur / Photo by Johannes Laurentius / Antikensammlung, Staatliche Museen, Berlin / Art Resource, NY.

PLATE 14 (FIG. 8.7). Detail, Pandora admired by gods and goddesses, on the red-figure calyx krater by the Niobid Painter, about 460 BC, inv. 1856,1213.1.

FIG. 9.1. Phalaris, the tyrant of Agrigentum, Sicily, burns the clever craftsman Perilaus in his own creation, the Brazen Bull. "Perillus condemned to the bronze bull by Phalaris," sixteenth-century woodcut by Pierre Woeiriot de Bouze. HIP / Art Resource, NY.

The horror of the Brazen Bull has a familiar ring, sounding mythic echoes from previous chapters. A hyperrealistic bull statue brings to mind the artificial cow created by Daedalus for Queen Pasiphae (chapter 4). Like Pasiphae's fake heifer, Phalaris's Brazen Bull was animated by the living human encased inside.[12]

Even more compelling mythic comparisons to the Brazen Bull would be the two deadly bronze automata created by Hephaestus for other powerful royal patrons. King Aeetes hoped to incinerate Jason with his awesome pair of fiery bronze bulls. And recall that King Minos's bronze automaton Talos could heat his body fiery-hot and crush victims to his chest, roasting them alive. Did the mythic parallels to Phalaris's bronze bull also occur to people in antiquity? In the absence of any surviving texts expressing direct links to the myths, that is unknowable but not implausible. Ancient tales and traditions about bronze bulls and heated metal statues were certainly pervasive in popular culture by the time of Phalaris.

Moreover, it turns out that artificial bulls were prominent talismans in the founding mother city of Phalaris's Acragas. Acragas was founded by colonists from Rhodes; Phalaris's father was born there. The island was well known for extraordinary feats of mechanical engineering, such as the Colossus of Rhodes (chapter 1). Evidence indicates that the complicated bronze astronomical calculating machine with thirty gears, the Antikythera mechanism, known as the world's first analogue computer, was made between the third and first centuries BC in Rhodes.[13] As we saw in chapter 5, Rhodes was also renowned for its animated bronze statues, celebrated in Pindar's poem (*Olympian* 7.50–54):

The animated figures stand
Adorning every public street
And seem to breathe in stone
Or move their marble feet

Among the wonders of Rhodes were two life-size bronze bulls. Were these bulls the prototypes for the Brazen Bull created for Phalaris of Acragas? The bronze bulls of Rhodes stood guard on the island's highest peak, Mount Atabyrios. (Guardians made of bronze were common in antiquity, chapter 1). We know that Phalaris was involved with the construction of the Temple of Zeus Atabyrios in Acragas, which was named for the mountain in Rhodes guarded by a pair of bronze bulls. But even more striking, the bulls of Rhodes were ingeniously manufactured to bellow. The bull sentries served as signal horns—they "bellowed loudly to warn the Rhodians of the approach of enemies."[14] A configuration of tubes in the bulls amplified the voices of human watchmen stationed on the mountain. It is not impossible that the Brazen Bull of Acragas was perversely designed with similar pipes to transform the victim's screams into bellowing sounds.

• • •

Signal horns and other megaphonic devices to augment the human voice were devised in various cultures of the ancient world. The artificial amplification of human voices to convey messages was attributed to Alexander the Great, who employed an enormous bronze horn or megaphone suspended on a large tripod to send signals in any direction to his army, several miles distant. The instrument was named after the prodigiously loud herald named Stentor in the mythic Trojan War (Homer *Iliad* 5.783). An exaggerated stentorophonic device also turns up in medieval legends about Alexander, whose phenomenally loud war trumpet, sometimes called the Horn of Themistius, could summon an army sixty miles away.[15]

More melodious mechanized sounds were also possible, emanating from a number of statues and automata, recalling the legendary singing maidens on the temple at Delphi (chapter 7). One example of a noisemaking statue is particularly appropriate here, namely, the statue of Athena created by the sculptor Demetrios (fourth century BC). According to Pliny (34.76) the statue was dubbed the "musical" or "bellowing" Athena (*musica* or *mycetica*—the manuscript is unclear). Strange sounds were said to emanate from the writhing serpents in the hair of the fierce Gorgon on the goddess's shield.[16]

A fascinating archaeological discovery in Cairo, Egypt (1936), reveals how some speaking and singing statues worked in antiquity. A large limestone bust of the sun god Ra-Harmakhis has a cavity in the back of the neck from which a narrow canal leads to an opening on the right jaw under the ear. The archaeologists speculate that a priest hiding behind the statue spoke into the cavity and tube, which modified his voice to make it seem that the god delivered oracles.[17]

• • •

A sublime song at dawn was said to issue from one of the Colossi of Memnon in Egypt, a pair of gigantic seated stone statues, sixty feet high, which were a tourist attraction in antiquity. Amenhotep III (Eighteenth Dynasty) erected the twin statues of himself in about 1350 BC at his temple on the Nile at Thebes. The Egyptians called the "singing" statue Amenophis, Phamenophes, or Sesostris; the Greeks called it Memnon. It was the northern statue—broken after the earthquake of 27 BC—that produced a marvelous tone or "voice" at dawn. In Greek myth, Memnon was the son of the goddess Eos and her undying mortal lover, Tithonus (chapter 3). As king of the Ethiopians, Memnon allied with the Trojans in the Trojan War. Some observers fancied that the speech or song uttered by Memnon's statue at sunrise was meant to console his mother, Eos, "Dawn." The rays of the sun made his eyes gleam, and the sound was heard "as soon as the sunbeam reached his lips." Visitors experienced the eerie sense that Memnon was on the verge of rising from his throne to greet the new day.[18]

The Roman historian Tacitus (*Annals* 2.61) noted that when struck by the sun's rays, Memnon "gives out the sound of a human voice, while the pyramids, made by the vast wealth of kings, loom like mountains in the impassable wastes of shifting sand." Some proposed that the sound was the result of the sudden expansion of the stone from the heat of the rays of the rising sun, perhaps activating internal levers that were attached to vibrating strings. (Perhaps a similar effect caused the Golden Charmers to "sing" at Delphi, chapter 7). Visiting the statues at sunrise in about 26 BC, the geographer Strabo and his friends (17.1.46) heard the sounds but could not be sure whether they came from the statue or from someone standing at the base. The main character in Lucian's satire *Philopseudes* (33; second century AD) claims to have heard a "prophecy" uttered by Memnon at dawn,

FIG. 9.2. The Colossi of Memnon, Thebes, Egypt, photo by Felix Bonfils, 1878. HIP / Art Resource, NY.

although "most visitors only hear unintelligible sounds." In AD 80–82, a Roman centurion named Lucius Tanicius inscribed the dates and times when he heard the song on thirteen visits. Many other ancient tourists left graffiti on the singing colossus—the last datable inscription is from AD 205. Some commentators maintained that after Emperor Septimius Severus restored the statue in AD 200, Memnon's song was never heard again, but the Christian Fathers Theodoret, Jerome, and others insisted that all the old Egyptian idols ceased to speak when Jesus was born.[19]

As we've seen, there were many ways to cause statues to appear to move, speak, or give the illusion of being alive.[20] Paul Craddock (an expert on ancient Near Eastern metallurgy) speculated that such "temple tricks" might have included making an idol that produced a tingling sensation when touched. Craddock's theory attempted to account for the enigmatic objects known as "Baghdad Batteries" discovered in 1936–38 in Iraq. The artifacts are thought to be either Parthian (ca. 250 BC to AD 240) or Sassanian (AD 224–640). The objects are controversial: some historians take them as evidence of early Persian experimentation with electricity. Unfortunately, the artifacts vanished in the looting of Baghdad's Iraq Museum in 2003, but written descriptions, diagrams, and photographs provide the details.

The small terra-cotta jars, each about five inches long, contain cylinders made of iron rods encased in rolled sheets of copper, sealed at the top with asphalt (bitumen) and at the bottom with a copper disc and asphalt: the copper-wrapped iron rod projects above the asphalt at the top. The jars' inner walls show evidence of corrosion. No wires were recovered: they may have been overlooked or corroded away. It is worth noting that very thin bronze "needles" have been found with similar jars (lacking cylinders) in the same region. The materials and construction seem to suggest a primitive galvanic cell. Modern experiments demonstrate that replicas of the Baghdad batteries produce a feeble 0.5 volt current, using a 5 percent electrolyte solution, with substances available in antiquity such as grape juice, vinegar, wine, or sulfuric or citric acid. If strung together and connected, a cluster of the jars might produce a higher output, enough to give a mild shock akin to static electricity.

The purpose of the cells is unknown; some suggest a medical function, while others propose a magical or ritual use. In Craddock's speculative scenario, if the jars were really electrical cells and were hidden and activated somehow inside a metal statue, the figure would seem charged with mysterious life and power. Anyone who touched it would be awed by a sensation of warmth, a strange buzzing vibration, and perhaps even a subliminal blue flash of light in a darkened chamber.[21]

● ● ●

Between the third century BC and the first century AD, fluctuating notes that imitated birdsong were made to issue from the beaks of realistically painted models of birds designed by Philo and Heron, famed inventors in Alexandria, Egypt, whose works are further described below. But even earlier, people were excited by an artificial bird capable of flight. This automaton was attributed to a philosopher-scientist-ruler named Archytas (ca. 420–350 BC), an associate of Plato. Archytas lived in Tarentum, a colony founded by Greeks in the heel of southern Italy.[22] Admired for his intelligence and virtue, Archytas was elected to the office of *strategos*, general, and he is thought to have influenced the idea of philosopher kings in Plato's *Republic*. Aristotle refers to Archytas's theories in several treatises, but Archytas's own writings no longer survive except in scraps.[23]

Horace addresses a poem to Archytas (*Ode* 1.28, "the Archytas ode"), and many ancient sources discuss Archytas, but a work by Aulus Gellius (writing in the second century AD) is the only extant text to describe the first self-propelled flying machine in the shape of a dove. What Archytas "devised and accomplished is marvelous" but not impossible, comments Aulus Gellius (*Attic Nights* 10.12.9–10). Aulus Gellius quotes "the philosopher Favorinus, a studious researcher of ancient records," who stated that Archytas "made a flying wooden model of a dove in accordance with mechanical principles." The Dove was "balanced with counterweights and moved by a current of air enclosed within it." The bird flew some distance, but "when alighted it could not take off again." Here, I'm sorry to report, the passage breaks off, and the rest of the text is lost.

Archytas's pathbreaking work on mechanical mathematics, cubes, and proportions allowed the creation of scale models. Much has been written by modern philosophers and historians of science on Archytas's principles of mechanics. The Dove appears to have been a plausible historical device. Mechanical engineers speculate that Archytas's Dove may have been tethered to a cord or stick and powered by steam or compressed air in a tube or metallic bladder controlled by a valve. It had to be reset after each flight (there is no evidence that the Dove had movable wings). A "reasonable reconstruction" of the Dove discussed by Carl Huffman in 2003 suggests that the bird was "connected by a string to a counterweight through a pulley" and its "motion was initiated by a puff of air that caused the dove to fly from a lower perch to an upper perch." Another hypothetical reconstruction, by Kostas Kotsanas, uses steam or compressed air to launch an aerodynamic bird.[24]

It is interesting to compare Archytas's Dove to two other historical mechanical devices from the fifth and fourth centuries BC, in the district of Elis in the Peloponnese, Greece, where the Olympic Games were held. The first mechanism featured a bronze eagle and dolphin. These figures were the moving parts of the ingenious starting gate for horse and chariot races in the Hippodrome at the Olympic Games. The eagle-and-dolphin mechanism was still operating in the second century AD, when Pausanias (6.20.10–14) described the starting gate. An official operated the machinery from an altar at the gate. To signal the start of the race, the eagle with outstretched wings suddenly flew up in the air and the dolphin leaped down, in view of the spectators. The device was originally made by the Athenian sculptor-inventor Cleoetas (480–440 BC) and later improved by Aristeides, a fourth-century BC artisan. Much admired for his hyper-realistic human statues with minute breathtaking details, such as inlaid silver fingernails, Cleoetas worked with the renowned Athenian sculptor Phidias to create the colossal gold and ivory statue of Zeus at Olympia in 432 BC (their workshop was discovered by archaeologists in the 1950s at Olympia; Phidias also created the enormous chryselephantine Athena statue in the Parthenon, chapter 8). It is likely that the eagle and dolphin on the starting gate were quite lifelike and, like Archytas's Dove, they must have been somehow tethered.

Elis also boasted a spectacle that took place during the Dionysia festival celebrating the god of wine. According to Pseudo-Aristotle (*On Marvelous Things Heard* 842A123), festival goers were invited into a building about a mile from the city to examine three large, empty copper cauldrons. When the people came out, the Elean officials then ostentatiously locked and sealed the building. After a while, the doors were unlocked and visitors allowed to reenter the building. They were surprised to find the three cauldrons now "magically" filled with wine. "The ceiling and walls appear to be intact, so that no one can discern any artifice." The trick apparently involved a hidden hydraulic technology of pumping the wine into the vessels. The date is unknown, but the description appears in a collection of notes gathered by Aristotle's students and followers.

As for Archytas, alongside his military, political, and scientific accomplishments in mathematics, geometry, harmonics, and mechanics, he was also credited—by Aristotle—with inventing a popular children's plaything, the clacking noisemaker known as the "clapper."[25] His toy clapper and his technological showpiece, the flying Dove, demonstrated mechanical principles while providing a delightful diversion—a welcome alternative to the cruel automata of other rulers.

• • •

A deceptively frivolous automaton of an invertebrate creature was constructed in Athens under oppressive Macedonian rule in the late fourth century BC. Demetrius of Phaleron was appointed to govern Athens by the Macedonian king Cassander in 317 BC. A well-educated orator who was a younger contemporary of Aristotle, Demetrius was sole ruler of Athens until he was forced into exile in 307 BC. He ended up in Alexandria, Egypt, where he was involved in establishing the great library and museum of Alexandria, where many inventors worked (see below). Demetrius later fell out of favor in Alexandria too, and was exiled to the hinterlands where he died of snakebite, about 280 BC.[26]

As tyrant of Athens, Demetrius was arrogant, given to excess and extravaganzas. Naturally, he despised democracy and he disenfranchised poor citizens. According to a lost history of the time by Demochares, a rival Athenian orator who defended democracy, in 308 BC Demetrius commissioned a moving replica of a giant land snail that "worked by some internal contrivance."[27] The Greek historian Polybius (12.13) tells us that this Great Snail led the traditional ceremonial procession of the Dionysia, Athens' great drama festival. Moving from Plato's Academy outside the city walls to the Theater of Dionysus, it traveled a distance of about 1.8 miles. The composition of the snail and its inner works are not detailed in Polybius's account. But the phrase "internal contrivance" suggests some self-propelling mechanism. In 1937, Alfred Rehm proposed that a man walking on a treadmill and another to steer were concealed inside the model of the large mollusk. Treadmills existed in antiquity; the massive, mobile "city-taker" siege machine, built in 323 BC by Posidonius for Alexander the Great, might have relied on a treadmill, and a Roman relief of the first century AD shows a huge construction crane powered by many men inside a large treadmill. But Rehm's theory is still debated.[28]

Why bother to create a gigantic moving replica of a lowly snail? One might note that the Dionysia festival was held in winter, when the rains begin and dormant land snails emerge in large numbers to crawl about, so real snails on the move would be conspicuous everywhere in Athens. Demetrius's oversized snail was so "realistic" that it even left a trail of slime as it inched along the route. This special effect would be easily achieved with a reservoir of olive oil released from a hidden pipe.

The most significant detail is the fact that the Great Snail was followed by a group of donkeys in the procession. This pairing of snail and asses would be part of the snide joke. Snails were proverbially slow, and because they carried their homes on their backs, they stood for impoverishment. Donkeys were associated with dull-witted, lazy slaves who work only when beaten.[29] As Demochares remarked (Polybius 12.13), the point of Demetrius's spectacle was to taunt "the slowness and stupidity of the Athenians." The Great Snail itself was harmless, but it was a dramatic and public way for the tyrant to humiliate the Athenians, whose democracy was being crushed by Macedonians and their collaborators.

• • •

A century later, in 207 BC, in Sparta, southern Greece, a malevolent dictator named Nabis seized power and ruled until 192 BC. His reign was long remembered for his barbarous acts, exiling, torturing, and killing masses of citizens. Nabis and his imperious wife, Apega (probably Apia, daughter of the tyrant of nearby Argos), collaborated to extort valuables and money from people under their rule. Their story is told by Polybius,

FIG. 9.3. Portrait of Nabis on silver coin, ruler of Sparta, 207–192 BC, inv. 1896,0601.49 © The Trustees of the British Museum.

a native of southern Greece who was born around the time of their overthrow. According to Polybius, Apega "far surpassed her husband in cruelty." When Nabis dispatched Apega to Argos to raise funds, for example, she would summon the women and children and then personally inflicted physical torture until they gave up their gold, jewels, and costly possessions (Polybius 13.6–8, 18.17).

As tyrant, Nabis welcomed a stream of nefarious characters, including pirates from Crete, to his kingdom.[30] Perhaps it was one of these opportunists who manufactured, on Nabis's orders, a mechanical Apega, a "machine" made to "resemble his wife with extraordinary fidelity" (Polybius 13.6-8, 16.13, 18.17). Inspired by his wife's deeds, "Nabis invented a female robot as evil and deceptive as Pandora," comments Sarah Pomeroy, a historian of Spartan women. The automaton was clothed in Apega's expensive finery. We can imagine that the artisan painted a plaster cast or wax model of Apega's own face to carry off the effect.

Nabis would summon wealthy citizens and ply them with wine while urging them to turn their property over to him. If any guest refused to comply, Nabis would say, "Perhaps my lady Apega will be more successful in persuading you." At the appearance of the replica of Apega, the inebriated guest would offer his hand to the seated "lady." She stood up, which triggered springs to raise her arms. Standing behind Apega, Nabis manipulated instruments in her back to cause her arms to suddenly clasp the victim. Working levers and ratchets, Nabis then tightened the false Apega's deadly embrace, drawing the victims closer by degrees. The fancy clothing hid the fact that the palms of her hands, her arms, and her breasts were studded with iron spikes, driven deeper into the victim's body by the increasing pressure. With this impaling device in the form of his wife, "Nabis destroyed a good number of men who refused his demands," wrote Polybius (13.6–8).[31]

By the time Nabis and Apega came to power, the late third century BC, many inventors and engineers in the Mediterranean world were already designing animated statues and other clever devices for peace and war. An example of a fourth-century apparatus, the ingenious *kleroterion* (a "randomization" device for selecting citizens to serve in civic offices) has survived. Along with the aforementioned Antikythera device, this lottery machine represents the tip of the iceberg; a great many other practical technological experimentations and other innovations have left no physical traces but were described in ancient texts.

By the fourth and early third centuries BC, military engineers in Italy, Carthage, and Greece had developed crossbow artillery and powerful torsion catapults, based on complex mechanical formulas and springs, for rulers such as Dionysius of Syracuse and Philip II of Macedonia. For his attempted conquest of Rhodes in 305 BC, Demetrius Poliorcetes, "Besieger of Cities," had his engineers construct the tallest mechanized siege tower ever built. Equipped with 16 heavy catapults and weighing about 160 tons, the iron-plated wooden "City Taker" required relays of more than 3,000 men to activate. Demetrius also deployed a gigantic battering ram manned by 1,000 soldiers. Archimedes of Syracuse is perhaps the most famous engineer of the Hellenistic era, devising numerous geometry theorems and designing a host of amazing machines utilizing levers, pulleys, screws, and differential gears, and ranging from astronomical apparatus and odometers to heat rays that ignited invading navies and the Claw, a massive grappling hook on a crane to grab and sink enemy ships.[32]

Given this rich legacy of classical and Hellenistic inventions, it seems safe to assume that Nabis's lethal Apega machine was modeled on technological precedents. The Apega replica was self-moving owing to springs that caused her to stand up and raise her arms; Nabis controlled the mechanisms to give the impression that the figure was operating under its own power. The Apega automaton was not heated but could kill victims by forcible embrace, recalling the way the mythical bronze robot Talos crushed people to his chest. Some historians have wondered whether the Apega device was an inspiration for the Iron Maiden, "Eiserne Jungfrau," the imaginary medieval torture/execution device, a metal cabinet shaped like a female with a spiked interior.

• • •

After the assassination of Julius Caesar in 44 BC, Rome was in turmoil. Marc Antony delivered the dramatic funeral oration over the bier in which Caesar's ravaged corpse lay out of sight. The historian Appian (*Civil Wars* 2.20.146–47) described the effects of the speech on the populace. Declaiming "in a kind of divine frenzy" and carried away by "extreme passion," Marc Antony grabbed a spear and with the point lifted the robe from Caesar's body and held it aloft so all could see the bloodstained cloth pierced with dagger thrusts. The mourners raised loud lamentations.

But the theatrical performance was not over. A hidden actor impersonating Caesar's voice recited the names of his murderers, further roiling the audience. Then from the coffin slowly rose the ravaged body of Caesar. It was an effigy made of wax, realistically displaying the twenty-three brutal knife wounds. The pièce de résistance followed, as the effigy rotated "by a mechanical device to display the pitiful sight." Crazed with rage and grief, the crowd rushed out to set fire to the Senate where Caesar was slain and tried to burn down the houses of the assassins. The sensational stagecraft of an automated, bloody, wax mannequin in Caesar's image was carefully orchestrated by Caesar's allies to manipulate the populace.

● ● ●

Some monarchs in the ancient Greco-Roman world were enthusiastic patrons of science and devised spectacles of animated statues in order to demonstrate their vast power and grandeur. Such wondrous machines told the world that the king could achieve the impossible.

One thwarted example of a Hellenistic ruler's attempt to glorify himself by means of a mechanized spectacle occurred during the reign of King Mithradates VI of Pontus, known for his prodigious ego and love of marvelous machines. Mithradates attracted the best craftsmen, scientists, and engineers to his court in the first century BC. His engineers built stupendous naval and siege machines, and the famous Antikythera device was looted from his kingdom by the Romans (70–60 BC). In about 87 BC, to celebrate his defeat of Roman forces in Greece, Mithradates commissioned a grandiose pageant. Bearing in mind classical Greek images of the winged goddess Nike hovering over victors' heads, the royal engineers created an immense statue of the goddess, suspended on cables out of sight. Similar deus ex machina technology was used on the stage in classical Greek theatrical performances, but this scheme was off the scale. At the climax of the festivities, the massive Winged Nike would dramatically descend, by a series of pulleys and levers, stretch out her hands and place a victor's crown on Mithradates's head, and then majestically ascend to the heavens. That was the plan. But the cables failed and Winged Victory smashed to the ground. The miracle was that no one was harmed, but the terrible omen was inescapable.[33]

• • •

A memorable, and in this case wildly successful, display of an autocrat's power took place in third-century BC Egypt, orchestrated by Ptolemy II Philadelphus (283–246 BC), of the powerful Hellenistic Macedonian Greek dynasty that ended with the famous queen Cleopatra in 30 BC. The Ptolemies were avid supporters of the arts and sciences at the new international research center in Alexandria, the library and museum complex founded in about 280 BC (it was mostly destroyed by fire in about 48 BC). Under the Ptolemies, Alexandria became the hub of scientific investigation, and the birthplace of machines, with mechanized public showpieces for theaters, processions, and temples, especially animated statues and automated devices.[34]

Ptolemy II Philadelphus married his sister, Arsinoe II, in 278 BC. As we saw, after her death he declared her a goddess and commissioned a miraculous floating statue of her (allegedly using magnets, chapter 5). But Ptolemy II's reign from 283 to 246 BC is most remembered for the outrageous splendor of his Grand Procession of 279/78 BC, a seemingly endless parade of exotic creatures, living tableaux, costumed dancers, and stunning automated displays that took place over several days. According to descriptions in a history of Alexandria by Callixenus of Rhodes (a contemporary of Ptolemy II who may have attended the event), the magnificent panorama included two dozen golden chariots drawn by elephants, followed by ostriches, panthers, lions, giraffes, and other animals, and a multitude of massive carts or floats, hundreds of performers dressed as satyrs and maenads and other mythic figures, larger-than-life realistic statues of divinities (including Alexander the Great), and engineering marvels. Sadly, like so many ancient texts crucial to our understanding of artificial life and automata in antiquity, Callixenus's works have vanished. But parts of his extensive account of the parade are preserved by the second-century AD author Athenaeus (*Learned Banquet* 5.196–203).[35]

Ptolemy's Grand Procession celebrated the Greek god of wine, Dionysus, and featured scenes from his mythology. Observers were dazzled by an enormous statue of Dionysus, 15 feet tall, holding out a huge golden goblet overflowing with wine, surrounded by a crowd of satyrs and Bacchantes, singers, and musicians. Another float bearing an immense winepress, about 30 feet long and 20 feet wide, was pulled by 300 men, while 60 men disguised as satyrs trampled the grapes. There was a vast wineskin made of leopard pelts borne on a heavy cart pulled by 600 men, while a continuous stream of wine poured out along the route. Yet another float featured two fountains gushing wine and milk (like those attributed to Hephaestus in Greek myth). The profusion of amazing and costly automated objects and statues on such a staggering scale evoked ancient versions of Uncanny Valley sensations. They fostered the illusion that all these constructions were being animated by the gods and goddesses themselves, giving the impression that Ptolemy could summon divine presences to celebrate his coronation.

After the cart carrying Dionysus, another astounding sight hove into view: a float with a gigantic seated female statue of Nysa, wearing a golden crown and draped in yellow-dyed garments covered in gold spangles. This Nysa was a true self-moving mechanical automaton. Periodically along the route Nysa stood up, poured a libation of milk from a golden *phiale*, and sat down again. She did this "without anyone putting their hands on the statue," commented Callixenus.

Who was Nysa? Nysa was the name of the mountain where the infant Dionysus was raised, nourished by rain nymphs. In the Hellenistic period, the mountain was personified as Nysa, Dionysus's nursemaid, so it was logical that she accompanied the god, dispensing milk.

The huge Nysa automaton, 12 feet high when seated, and the large reservoir for milk would have been heavy. Indeed, Nysa's cart was reportedly 12 feet wide and pulled by 60 men. Like the other oversized statues, Nysa was not bronze or marble but fabricated of terra-cotta, wood, plaster, and wax and realistically painted. To operate faultlessly and in a dignified manner for the entire length of the slow-moving procession (estimated to have been about 3 miles long), the automaton mechanism, as modern engineers agree, must have been technologically robust.

How did the Nysa automaton work? In 2015, historians of mechanical engineering Teun Koetsier and Hanfried Kerle analyzed and diagrammed several hypothetical designs. If the statue was 12 feet high when sitting, it would have been 15 feet tall when standing. Assuming it was powered by mechanical means and with components available at the time, they conclude that a complex arrangement of cams, weights, and a sprocket chain or gear wheels were carefully timed to make Nysa rise from her chair, pour milk, and sit down in a slow, stately manner.

Who made the unprecedented Nysa automaton, one of world's first working robots? The ancient sources do not say. One candidate was the engineer Ctesibius, thought to have been the first director of the museum at Alexandria. No writings by Ctesibius survive, sad to say, but his inventions, based on hydraulics (pumps, siphons) and pneumatics (compressed air), were very highly regarded, described by Vitruvius, Pliny, Athenaeus, Philo of Byzantium (who worked in Alexandria), Proclus, and Heron of Alexandria. Ctesibius was active in 285–222 BC, and he created a pneumatic drinking horn in a temple honoring Ptolemy II's late wife, Arsinoe II. Ctesibius, or some of his colleagues, would seem to be the most likely builders of the Nysa robot in Ptolemy's Grand Procession.[36]

What about Philo of Byzantium (Philo Mechanicus), the eminent Greek engineer and writer who lived most of his life in Rhodes and Alexandria? His exact dates are unknown, but it is now believed that Philo was born about 280 BC, making him a bit too late for Ptolemy II's Grand Procession. Philo's impressive list of machines and plans for self-moving devices in the forms of humans and animals were greatly admired in antiquity and the Middle Ages and are still studied today.[37]

• • •

Philo's compendium of mechanical works ranged from siege towers to theatrical machines, and he designed a host of devices and automata. Most of his treatises have been lost, but the plans and instructions were preserved in later sources, by Heron and Islamic writers.[38] We've already met Philo's version of the god Hephaestus's robotic assistants, a realistic life-size serving maid with the ability to pour a cup of wine and then dilute it with water (chapter 7). That self-moving mechanical woman of the third century BC has been hailed as the first man-made "robot," although the Nysa automaton preceded her by some years. Philo preferred to make cunning miniature mechanisms, all the more astounding because of their small scale.

One of Philo's pieces features an artificial bird that chirps when an owl turns to face it and falls silent when the owl turns away. The mechanism depends on water poured into a vessel to displace air, which is forced out through a small pipe to the bird's beak; oscillating wavelengths produce notes with different frequencies. A rotating shaft controlled by the water level causes the owl's rotation. Philo also designed a bird that raises its wings in alarm as a snake approaches its nest. Pouring water into a reservoir lifts a float connected by a rod to the bird's wings. Yet another enchanting automaton depicts a dragon that roars when a figure of Pan faces it, and relaxes when Pan turns away (a variant features a deer drinking while Pan is turned away).[39]

Philo was a strong influence on another leading Alexandrian inventor, Heron of Alexandria (AD 10–70), many of whose writings and designs for engines, machines, and automata still survive. Heron assembled amazing machines enacting charming mythic vignettes, using hydraulics and other mechanisms to make them move in complex ways. He also created "Dionysian" devices that appeared to produce wine spontaneously, recalling the self-filling cauldrons in Elis and the wondrous spectacles in Ptolemy's Grand Procession, described above. Heron famously advised fellow engineers to make small automata so that no one could suspect that they were worked by a person hidden inside. In his treatises *On Making Automata* and *Pneumatica* Heron describes stationary and moving devices with complex forms of motion, including "snake-like" movements. His instructions and specifications permit engineering technicians to construct working models.[40]

A typical assemblage designed by Heron features a bronze Heracles shooting an arrow at a bronze serpent that hisses when struck. Heron also devised miniature automatic theaters. The theater rolled onto a stage by itself, stopped, and performed with "fires flaring on altars, sound effects, and little dancing statues"; then it rolled offstage. It has been called the first programmable device.[41] To initiate the chain reactions that create a series of sights and sounds on the little stage, the operator simply pulls a string to activate a steadily descending lead weight in a sand *clepsydra* (a mechanism based on liquid or sand draining at a steady pace) and then steps aside as spectators observe the spellbinding show (see fig. 9.4 for a working replica of the theater). The stage doors automatically open and close on five scenes of a little Trojan War tragedy titled *Nauplius*. First, shipbuilders are seen and heard hammering and sawing wood. Next the men push the ships into the sea. Now rocking ships sail on a rough sea with leaping dolphins. A torch signal lures the ships to a rocky shore at night, and in the last act the Greek hero Ajax is seen swimming amid wrecked ships while Athena appears on the left and disappears stage right. Suddenly lightning strikes Ajax and he vanishes in the waves.[42]

These exquisitely constructed mechanical dramas made by Philo and Heron reproduced in reality some of the phantasmagoric imaginary panoramas on Pandora's golden crown and Achilles's shield made by Hephaestus. As described in Homer's *Iliad* and *Odyssey*, the god constructed lifelike miniature people and creatures that seemed to move and make sounds (chapters 5, 7, 8).

• • •

Many of the designs for automata devised by Philo and Heron were preserved in early medieval Arabic and Islamic texts—for example, by the Banu Musa brothers in Baghdad (ninth century AD, Iraq) and al-Jazari in the twelfth century. These Hellenistic and medieval Near Eastern influences on European automata and machines of the Middle Ages have been extensively studied.[43] Mechanical innovations in early China are also well documented by historians. By the third century BC in China, for example, Qin dynasty (221–206 BC) artisans had developed mechanized puppets and other devices. In about AD 250, the engineer Ma Jun invented a precise south-pointing figure in a gear-driven chariot and a puppet theater powered by a waterwheel.[44]

During the Tang dynasty (AD 618–907), technological advances resulted in a profusion of sophisticated automata and self-operating devices. Typical examples include an iron mountain with hydraulic pumps to spew liquor from a dragon's mouth into a goblet and a fleet of moving boats with automated servants to pour wine. Tang engineers created many automatic devices for Empress Wu Zetian (r. AD 683–704). A Buddhist convert,

FIG. 9.4. Replica of the automated Theater of Heron of Alexandria, based on Philo's designs. Top, the theater doors open to reveal the sights and sounds of shipbuilders hammering and sawing, controlled by inner workings. Center, ships rock on the choppy sea with leaping dolphins. Next, Ajax drowning amid the wrecked ships, with Athena moving in the foreground. Bottom, mechanism for moving Athena. Working model constructed by Kostas Kotsanas, courtesy of the Kotsanas Museum of Ancient Greek Technology.

Empress Wu sought to emulate and surpass the veneration of Buddha's relics in India by King Asoka, the great ruler of the Mauryan Empire in the third century BC. Many legends had grown up around Asoka and were brought back to China by Chinese Buddhist pilgrims. One of the most intriguing legends about Asoka involves mechanical beings.[45]

• • •

Robotic guardians appear in Buddhist legends set in India during the time of the historical kings Ajatasatru and Asoka. Both rulers were entrusted with safeguarding the precious relics of Buddha, whose death occurred sometime between 483 and 400 BC. The Indian legends are remarkable, not only because they describe mechanical warriors defending the bodily remains of Buddha, but because the stories explicitly link the robots to automata invented in the Hellenistic Greco-Roman world. This unexpected historical and geographical connection invites deeper investigation.

King Ajatasatru of Magadha (northeastern India) reigned from about 492 to 460 BC, in his fortified capital of Pataliputta (the city's ruins lie under modern Patna). According to Buddhist traditions, he met Buddha and became his devotee. After Buddha's death and cremation, Ajatasatru constructed a vast *stupa* (dome) over a deep underground chamber containing the holy ashes and bones. Then, it is said, Ajatasatru devised special defenses to protect Buddha's relics. Traditional Hindu and Buddhist architecture featured armed guardians of doors and treasures (*dvarapalas* and *yakshas*), sometimes sculpted in the form of giant warriors (fig. 9.5).

But Ajatasatru's guardians were extraordinary. He had his engineers in Pataliputta make a set of automaton warriors to defend the remains of Buddha. It is worth mentioning that according to ancient Jain texts Ajatasatru deployed novel military inventions: examples include a powerful catapult that hurled massive boulders and a mechanized, heavily armored war chariot, something like a "tank" or "robot," which wielded whirling maces or blades. His automaton guards were also said to have whirling blades.[46]

The legend relates that it was predestined that Ajatasatru's automaton guards would remain on duty until a future ruler—King Asoka—would discover and disable the robots, gather up the sacred relics of Buddha,

FIG. 9.5. Two traditional *dvarapala-yaksha* guardian warriors armed with spears on either side of a table holding Buddha's relics, panel relief, Kushan, Gandhara, Swat, first to second century AD, inv. 1966,1017.1 © The Trustees of the British Museum. The panel relief is flanked by a pair of six-foot-tall guardian warriors, found at ancient Pataliputta, Mauryan Empire, third to first century BC, plate 13, E. J. Rapson, *Cambridge History of India* (1922). Collage by Michele Angel.

and distribute them among tens of thousands of shrines throughout the realm. King Asoka (304–232 BC) ruled the powerful Mauryan Empire from about 273 to 232 BC in Pataliputta and became a follower of Buddha. During his long reign, Asoka constructed many *stupas* to enshrine a multitude of Buddha's relics across his vast kingdom, fulfilling the prophecy of Ajatasatru.[47]

Several Hindu and Buddhist texts in various translations describe Ajatasatru's automaton warriors guarding the relics until the arrival of Asoka. The wooden androids were said to whirl with the speed of the wind, slashing intruders with swords. Some traditions attribute their creation to Hindu divinities: Visvakarman, the engineer god, or Indra, the guardian god. But the most arresting and mysterious account of the robot guards has come down to us through a tangled route: it appears in the collection of tales known as the *Lokapannatti* from Burma, a Pali (sacred language) translation of an older, lost Sanskrit text, which is itself known only from a Chinese translation. The dating of the *Lokapannatti* is uncertain, perhaps eleventh or twelfth century, but the stories "drew on a rich store of 'legends' about Asoka," a "large variety" of much older oral traditions and lost texts.[48]

The tale recounts that many *yantakara* (robot makers) lived in the land of the *Yavanas* (Greek-speakers; people of the West) in *Romavisaya*, the "kingdom of Rome," a generic term for the West, namely, Greco-Roman-Byzantine culture. The *Yavanas*' secret technology of robots (*bhuta vahana yanta*, "spirit movement machines") was closely guarded by their government. In "Rome," robots carry out trade and farming, and they capture and execute criminals. No robot makers are ever allowed to leave "Rome" or reveal their secrets—if they do, robot assassins will pursue and kill them. Rumors of the fabulous Roman robots reached India, inspiring a young artisan-engineer who wished to learn how to make automata. The young man lived in Pataliputta. As noted above, Pataliputta was the large fortified city built by King Ajatasatru in about 490 BC. It reached a peak of prosperity as King Asoka's capital in the mid-third century BC.

By magical plot contrivances the young man of Pataliputta fulfills his vow to be reincarnated in "Rome"—the Greek-influenced West. He marries and has a son with the daughter of the master robot engineer in Rome. He learns the robot maker's craft. Then he steals the plans for making robots, sews the papyrus under his skin, and departs for India. Knowing that he will be killed by pursuing robot assassins before he can reach India, he has already instructed his son to take his corpse back to Pataliputta. His son does so, and retrieves the plans. The son creates an army of automated soldiers for King Ajatasatru to protect Buddha's relics hidden in a deep underground chamber of the secret *stupa*.

The hiding place and the robots fall into long obscurity. Then one day Ajatasatru's descendant, the great emperor Asoka, hears the story of Buddha's hidden relics and the prophecy. Asoka searches everywhere until he discovers the *stupa* with the underground chamber guarded by the fierce android warriors. In the meantime, the Roman emperor learns of the theft of Western technology: Why, he wonders, does the secret technology in India so closely resemble our own? The Roman emperor sends a gift containing a robot assassin to kill Asoka, but it is thwarted. Violent battles ensue between Asoka and the automaton guards in the underground chamber. Finally, Asoka locates the miraculously long-lived engineer's son, who shows him how to dismantle and control the "Roman" robots. Emperor Asoka now commands a large robot army himself.

In some versions, the whirling guardian automata are driven by a waterwheel or some other mechanism. In one tale, the engineer god Visvakarman helps Asoka, destroying the robots by shooting arrows precisely into the bolts that hold the spinning constructions together.[49] The motif of cleverly disabling the mechanical guardians calls to mind the techno-witch Medea's destruction of the bronze robot Talos, when he threatened to kill Jason and the Argonauts, by removing the crucial bolt in his ankle (chapter 1).

The "science-fiction" saga of the Roman robots guarding Buddha's relics highlights the fear of losing control of artificial beings, an age-old theme that appeared in the Greek myth of the sown dragon-teeth army (chapter 4). "Robots can turn on their makers and kill them," notes Signe Cohen in her study of ancient Indian automata. But the story raises more challenging questions. "Did such technology," she asks, "really exist or are these stories simply religious myths and folktales?"[50]

The story clearly relates the mechanical beings defending Buddha's relics to advanced automata inventions that originated in *Roma-visaya*, the Greco-Roman West. These narratives, remarks Daud Ali, seem to "encode, albeit obliquely, the real movement and circulation of cultures of 'techne,' including both real and imagined objects," between India and the West.[51] How ancient is this kernel of historical reality in the lost Sanskrit tale included in the *Lokapannatti*? Were the legendary robot guardians in the *stupa* modeled solely on working automata created in the late Byzantine or medieval Islamic and European periods, as scholars generally assume? Or is it possible that oral lore about the robot guards could have arisen even earlier, influenced by Indian knowledge of real Hellenistic mechanical marvels like those created in Ptolemaic Alexandria in the third century BC, the time frame of the Asoka story?

The historical setting of the tale points to technological exchange about automata between the Mauryan emperors of India and Hellenistic kings. Evidence from history and archaeology confirms cultural contact by the fifth and fourth centuries BC. Notably, the ancient Jain texts, mentioned above, reported that King Ajatasatru's engineers were constructing military machines in the fifth century BC. Greco-Buddhist syncretism and mutual influence in philosophy and art intensified after Alexander the Great's campaigns in what is now Afghanistan, Pakistan, and northern India.[52] We know that around 300 BC, the two Greek ambassadors, Megasthenes and Deimachus, arrived in the Mauryan court, and they resided in Pataliputta—a city with outstanding Greek-influenced art and architecture. Pataliputta, we recall, was the hometown of the engineer who obtained the plans for making robots from "Rome."[53]

King Asoka lived in the third century BC, at a time when automata and other devices were proliferating in Alexandria and other centers of technology in the West. Throughout his kingdom, Asoka left many inscribed pillars and rock inscriptions, some written in ancient Greek and others referring to Hellenistic kings by name, attesting to ongoing cultural exchange and trade with the West. Asoka sent emissaries and corresponded with several Hellenistic rulers, including Ptolemy II Philadelphus in Alexandria, whose spectacular procession in 279/78 BC featured marvelous displays of robotic mythic figures like Dionysus and Nysa. Asoka's envoys came to Alexandria, and Ptolemy II sent his own ambassador, a Greek named Dionysius, to Asoka's court in Pataliputta.[54]

Further evidence of long-lasting cross-cultural influence comes from the journal of the Chinese monk Fa Hsien, one of many Buddhist pilgrims who traveled to Pataliputta, Asoka's city, in about AD 400. Fa Hsien witnessed the traditional annual procession celebrating Buddha, presumably begun in Asoka's day. The monk describes the magnificent parade of large four-wheeled carts bearing colossal structures, imposing replicas of *stupas* five stories high, a succession of towering images of Buddha, Bodisattvas, and other divine beings of gold, silver, and lapis lazuli, with colorful silk banners and canopies, attended by hosts of singers, dancers, and musicians. Fa Hsien does not mention mechanized statues (although automated Buddhist figures were displayed in parades in China in this era).[55] One has a sensation of déjà vu, so closely does the scene in Pataliputta resemble the Grand Procession of Ptolemy II Philadelphus in Alexandria in 279 BC, a half century earlier.

Was the tale of Asoka and the robots known to Empress Wu (b. AD 624) and her engineers in Tang China? There were many real and imaginary automata in her era. A large golden Buddha surrounded by rotating mechanical attendants that periodically bowed and tossed incense had been created by the engineers Xie Fei and Wei Mengbian for processions in about AD 340. A sixth-century AD Chinese story recounts how workmen ordered to destroy two Buddha statues were attacked by wrathful Vajrapani guardians. Empress Wu knew the monk Daoxuan (AD 596–667) who designed sacred technology for shrines; in his writings Daoxuan described a fantastic Buddhist monastery in India with many automaton guardians in human and animal forms. We know that Empress Wu idolized Asoka, and that her engineers built "celestial" buildings for Buddha's relics, as well as mechanical marvels. It seems possible that the Chinese monks who transported Buddha's teachings, relics, and *stupa* designs from India to China also transmitted the legend of Asoka and the robots—a story that is preserved in a Chinese translation.[56]

IMAGINING ANCIENT ROBOTS

How might we moderns imagine Emperor Asoka's encounter with ancient "Roman robots"? How were the automatons guarding Buddha's relics visualized when the tale was told in antiquity? Traditional guardian *dvarapala* and *yaksha* statues defended Buddhist stupas and shrines from the Mauryan Empire period. These were warrior figures wielding bows, maces, and swords, sometimes monumental (fig. 9.5). But no ancient illustrations of the legendary self-moving guardians of Buddha's relics have been identified.

In Buddhist legends and artworks, the Buddha, his teachings, and his physical relics are protected by Vajrapani, the fierce bodhisattva armed with a lightning bolt. Remarkably, some of the earliest sculptural images of Buddha in Gandharan-style art of northern India (first century BC to seventh century AD) show Buddha in classical Greco-Roman garb and guarded by Heracles, the hero of classical myth. As Heracles merged with the persona of Vajrapani, the muscular, bearded guardian was shown wearing the Greek strongman's signature lion-skin cape, and his club is transformed into Vajrapani's distinctive *vajra*, the lightning bolt (fig. 9.6). Some reliefs show Heracles-Vajrapani carrying a sword, the weapon said to be wielded by the robots in the *Lokapannatti* story.[57] The artistic syncretism that merges the Greco-Roman mythic figure of Heracles with Vajrapani as a defender of Buddha chimes with the Buddhist story that Greco-Roman-style robots served as guardians for Buddha's relics. One might speculate that the automaton warriors defending the relics in the *stupa* might have been imagined as figures that combined classical Greek and Indian features.

FIG. 9.6. Buddha guarded by Heracles/Vajrapani, panel relief, Kushan, Gandhara, Pakistan, second to third century AD, inv. 1970,0718.1. © The Trustees of the British Museum.

The Arhats (Chinese Luohan), four original disciples of Buddha, were charged with defending the faith in early Indian sutras. Later in China, their number rose to eighteen. The earliest known artistic impressions of the Luohans (ninth century AD) depicted them as non-Chinese foreigners from the West. Although no link between the Luohans and the story of the "Roman" robots that defended Buddha's relics has been identified, at some point the Luohans were imagined as fierce bronze automata with fighting skills. The theme appears in the Shaolin kung fu movie *18 Bronzemen* (Joseph Kuo, 1976), set in the Qing Empire.

The fantasy of discovering long-forgotten automaton technology from some archaic civilization views robot technology with a mythological sensibility and lens. Notably, Hesiod suggested that the bronze robot Talos was of an earlier age. The notion of "ancient robots" has become a popular science-fiction theme. In 1958, the fantastical Buddha Park sculpture garden, Xieng Kuan near Vientiane, Laos, was created. The park is populated with colossal Hindu-Buddhist guardian statues (fig. 9.7), some of which resemble vintage robots. Made of concrete, they are deliberately designed to look like weathered antiquities. Meanwhile, in Japan, robots both imaginary and real were embraced with alacrity after World War II, a cultural feature that some attribute to Buddhist spirituality. Masahiro Mori, a devout Buddhist, not only was the first to articulate the Uncanny Valley effect; he also believed that robots could even have a "Buddhist nature." In some forms of Japanese and Chinese Buddhism, moreover, there is no

FIG. 9.7. Imaginary robot-like Buddhist guardians, created in 1958 to look ancient, Buddha Park, near Vientiane, Laos. Left, photo Kerry Dunstone; right, photo Robert Harding; Alamy Stock.

bright line between original and replica, essence and copy.[58]

Popular Japanese manga and anime artistic and literary forms arose after World War II and often featured artificial beings and robots. Notably, the anime-manga series *Mazinger Z* (1972–74; *Tranzor Z* in the United States) describes a super-robot modeled on ancient Talos-type steel prototype automata excavated by archaeologists on a Greek island loosely based on Rhodes. The conceit is that an ancient lost civilization, the "Mycene Empire," deployed these remote-controlled robots in battles. Another more recent example is the anime film *Laputa: Castle in the Sky* (1986, Hayao Miyazaki for Studio Ghibli, Tokyo). Drawing on ancient Hindu epics, the story involves the revival and dismantling of long-lost robot guardians created by a vanished culture. An international group of retrofuturists, mecha artists, and robot model makers fabricate intricate replicas of "abandoned" robots, cast as survivors of antiquity unearthed in archaeological ruins. A typical example is "Whistlefax." According to his fictional backstory, he arose from "the wastes of a world racked by violence," the devastated "ruins of a once great civilization overrun by hordes of haunted robots. Possessed by the souls of angry soldiers, these rusted hulks of an age gone by are given a new purpose, to punish those who plunged the world into conflict without purpose but to the profit of the few."[59]

When did the Buddhist tale of Asoka and the "Roman robots" first arise in India? The narrative appears to reflect genuine knowledge of actual engineering feats in the historical period of Ptolemy and Asoka, by the third century BC. We know that the Mauryan and Hellenistic courts sent envoys to each other, and they exchanged luxurious gifts to show off their cultural achievements. Note that the legend relates that plans for making automata reached India, and the emperor of the Greco-Roman West sent a gift box containing a robot to Asoka. One cannot hope to pinpoint the original date of the legend. But it seems safe to assume that Asoka and his contemporaries would have been familiar with—and perhaps even observed plans or miniature scale models of—automata and other mechanical marvels in the West.

• • •

Mechanical devices and automata in mythology and in real life provoked questions about ontology, humans and nonhumans, nature and artifice; they challenged the borders separating illusion, reality, and possibility. A large group of myths show that animated statues were certainly conceivable at a very early date, long before historical mechanical devices proved that imitating life with technology was practical. "Ancient mechanics *surprised* its audience," remarks Sylvia Berryman, and "experience with technology changed views about what results could be produced," about what might be possible. Human imagination and curiosity drive creativity and innovation.[60] Mythological stories about artificial life and as-yet-unknown technology can be considered another, valid kind of "experience." Imaginative scenarios in myth might well have helped shape ancient ideas and speculations about what results *might* be produced, what wonders *might* be possible, if only one possessed the radically superior technology and expertise of a Daedalus, Prometheus, or Hephaestus.

Were some marvels of artificially created life in the mythic traditions cultural fantasies that embellished and extrapolated real-life theories of technology or actual—if simpler—technological experiments? Or, conversely—just as modern science fiction can anticipate future scientific discoveries and sometimes even inspire technological innovations—is it possible that tales of divine and legendary automata and devices challenged and inspired living inventors to design self-moving objects and machines? Were mythic narratives and scientific imagination interrelated? The AI historian and futurist George Zarkadakis considers the links between old stories about robots and AI research. He proposes a feedback loop, a coevolution between mythic narratives and "scientific endeavors throughout history."[61] Speculations about original influence are impossible to resolve. But one can discern mythical chords within some historical inventions in antiquity. Indeed, it is striking that, just as ancient mythology about artificial life and self-moving devices imagined technological wonders made by divine craftsmen, so many historical inventors crafted automata and mechanisms to illustrate or evoke the ancient myths.

Millennia ago, visionaries initiated a series of "science-fiction" thought experiments about superior beings creating artificial life, expressed in mythical language. These imaginary automata, especially those like Talos and Pandora, with physically realistic forms and quasi-conscious "minds" that could interact with human beings on earth, evoked ambivalent reactions of awe, hope, and terror. Later, a group of brilliant inventors constructed real automata and self-moving devices that replicated natural forms, and their speculations and designs stimulated further experiments and innovations. As in the world of mythology, real automata and machines could be used to dazzle, deceive, and dominate. As we saw in chapter 8, inherent in the Pandora myth and proclaimed in Sophocles's paean to human ingenuity, *techne*, and ambition is a clear warning that these gifts can lead humans to glory or to evil.

The exciting dream of artificial life, first spun in storytelling imaginations, began to be realized in technological designs and engineered machines in antiquity. The next two millennia witnessed immense technological change. Yet by the end of the twentieth century, the journey of human creative vision and innovation had really only just begun. Advances are now accumulating at warp speed. Suspended above the uncanny abyss of replicating life itself, we still swing between hope and terror unleashed by humans' insatiable quest to imitate and improve nature.

EPILOGUE

AWE, DREAD, HOPE

DEEP LEARNING AND ANCIENT STORIES

Ancient myths articulated timeless hopes and fears
about artificial life, human limits, and immortality.
What could we—and Artificial Intelligence—learn from the classical tales?

THE MIX OF exuberance and anxiety aroused by a blurring of the lines between nature and machines might seem a uniquely modern response to the juggernaut of scientific progress in the age of technology. But the hope—and trepidation—surrounding the idea of artificial life surfaced thousands of years ago in the ancient Greek world. Imaginative myths expressed and struggled with the awe, dread, and hope summoned by the creation of animated statues, attempts to surpass human limits, and the pursuit of immortality. This is a discussion one might say that the ancient Greeks began.[1]

The question of what it meant to be human obsessed the ancient Greeks. Time and again, their stories explore the promises and perils of staving off age and death, enhancing mortals' capabilities, replicating nature. The complex network of myths about Prometheus, Jason and the Argonauts, Medea, Daedalus, Hephaestus, Talos, and Pandora—all raised basic questions about the boundaries between biological and manufactured beings.

The most enduring, best-loved Greek myths—along with many other long-forgotten ancient tales—spin thrilling adventures well worth knowing for their own sake. But when we recognize the old stories as inquiries into *biotechne* (*bios*, life; *techne*, craft), these "science fictions" of antiquity take on new significance. Deeply imbued with metaphysical insight and forebodings about divine and human manipulation of natural life, the mythical stories seem startlingly of our moment.

The fantasies of imitating and augmenting life inspired haunting dramatic performances on the stage and indelible illustrations in classical vase paintings, sculpture, and other artworks. Meanwhile, in about 400 BC the philosopher-engineer Archytas caused a sensation with the first mechanical bird in flight. By the Hellenistic era, Heron of Alexandria and other brilliant engineers were devising a multitude of automated machines driven by hydraulics and pneumatics. The Greeks recognized that automata and other artifices in natural forms—whether imagined or actual—could be either harmless or dangerous, and they could be used for work, sex, spectacle, or religion, or to inflict pain or death. Clearly, *biotechne*, both real and imaginary, fascinated the ancients.

Taken together, the myths, legends, and lore of past cultures about automata, robots, replicants, animated statues, extended human powers, self-moving machines, and other artificial beings, and the authentic technological wonders that followed, constitute a virtual library and museum of ancient wisdom and experiments in thinking, a priceless resource for understanding the fundamental challenges of biotechnology and synthetic life on the brink today. A goal of this book has been to suggest that on deeper levels the ancient myths about artificial life can provide a context for the exponential developments in artificial life and Artificial Intelligence—and the looming practical and moral implications. I hope that rereading those ancient stories might enrich today's discussions of robotics, driverless cars, biotechnology, AI, machine learning, and other innovations.

We saw how the god Hephaestus made a fleet of "driverless" tripods that responded to commands to deliver food and wine. Even more remarkable was the covey of life-size golden female robots he devised to do his bidding. According to Homer, these divine servants were in every way "like real young women, with sense and reason, strength, even voices, and they were endowed with all the learning of immortals." More than twenty-five hundred years later, Artificial Intelligence developers still aspire to achieve what the ancient Greeks imagined that their god of technological invention was capable of creating.

Hephaestus's marvels were envisioned by an ancient society not usually considered technologically advanced. Feats of *biotechne* were dreamed up by a culture that existed millennia before the advent of robots that win complex games, hold conversations, analyze massive mega-data, and infer human desires. But the big questions are as ancient as myth: Whose desires will AI robots reflect? From whom will they learn?

In 2016, an experiment in AI machine learning became a cautionary tale, when Microsoft invented the teenage fem-chatbot Tay. Intricately programmed to mimic neural networks in the human brain, Tay was supposed to learn from her human "friends" on the social network Twitter. She was expected to articulate conversational gambits without filters or behavioral supervision. Within hours of Tay's going live on Twitter, malicious followers conspired to cause the bot to morph into a tweeting troll spewing racist and sexist vitriol. Within days, Tay was terminated by her makers. Her easily corrupted learning system dampened optimism about self-educating AI and smart robots, but only momentarily. Tay's replacement, Zo (2107) was supposedly programmed to avoid chatting about religion and politics, but she too went rogue on Twitter.[2]

In Greek myth, the capstone of Hephaestus's divine laboratory was the female android commissioned by Zeus. To punish humans for accepting the technology of fire stolen by Prometheus, Zeus commanded Hephaestus to fabricate Pandora (chapter 8). Each of the gods endowed the artificial maiden with a human trait: beauty, charm, knowledge of the arts, and a deceitful nature. As the vengeful god's AI agent, Pandora executed her mission to unseal a jar of disasters to plague humankind forever. She was presented as a wife to Epimetheus, a man known for his impulsive optimism. As we saw, Prometheus warned humankind that Pandora's jar should never be opened. Are Stephen Hawking, Elon Musk, Bill Gates, and other prescient thinkers the Promethean Titans of our era? They have warned scientists to halt or at least slow the reckless pursuit of AI, because they foresee that once it is set in motion, humans will be unable to control it. "Deep learning" algorithms allow AI computers to extract patterns from vast data, extrapolate to novel situations, and decide on actions with no human guidance. Inevitably AI entities will ask—and answer—questions of their own devising. Computers have already developed altruism and deceit on their own. Will AI become curious to discover hidden knowledge and make decisions by its own logic? Will those decisions be ethical in our human sense? Or will AI's ethics be something "beyond human?"

Released from Pandora's jar—much like the computer viruses let loose by a sinister hacker who seeks to make the world more chaotic—misfortune and evil flew out to prey upon humans for as long as the world exists. In simplistic fairy-tale versions of the myth, the last thing to flutter out of Pandora's box was *hope*. But in darker versions, the last thing in the jar was "anticipation of misfortune." And Zeus had programmed Pandora to slam down the lid, trapping foreknowledge inside. Deprived of the ability to anticipate the future, humankind was left with what we call "hope." As was true of Epimetheus, foresight is not our strong point.

Yet foresight is crucial as human ingenuity, curiosity, and audacity continue to push the frontiers of biological life and death and the melding of human and machine. Our world is, of course, unprecedented in the scale of techno-possibilities. But that unsettling oscillation between techno-nightmares and grand futuristic dreams—that is timeless. The ancient Greeks understood that the quintessential attribute of humankind is always to be tempted to reach "beyond human," and to neglect to envision consequences. We mirror Epimetheus, who accepted the gift of Pandora and only later realized his error.

In 2016, Ray Crowder, an engineer at Raytheon, created three miniature learning robots. He gave the robots classical names: Zeus, Athena, and Hercules. With neural systems modeled on those of cockroaches and octopuses, the little solar-powered robots were endowed with three gifts: the ability to move, a craving for darkness, and the capacity to recharge in sunlight. The robots quickly learned to travel and soon understood they must venture into excruciating light in order to recharge or die. This seemingly simple learning conflict of these creatures that were *made, not born*, parallels human "cognitive economy," in which emotions help the brain allocate resources and strategize. Other AI experiments are teaching computers how human strangers convey goodwill to one another, and how mortals react to negative and positive emotions.[3]

Since Hawking warned that "AI could spell the end of the human race," some scientists are proposing that we could teach human values and ethics to robots by having them read stories. "Fables, novels, and other literature," even a database of Hollywood movie plots, could serve as a kind of "human user manual" for AI computers. One such system is named Scheherazade, in homage to the heroine of *The One Thousand and One Nights*. Scheherazade was the legendary Persian philosopher-storyteller who had memorized myriad tales from lost civilizations. She saved her own life by reciting these enchanting stories to her murderous captor, the king. The first stories uploaded into the Scheherazade AI were simple narratives that show computers examples of how to behave like good rather than psychotic humans. With the goal of empathetic interactions with human beings and appropriate responses to their emotions, more complex narratives would be added to the computer's repertoire. The idea is that stories would be valuable when AI entities achieve the human mental tool of "transfer learning," symbolic reasoning by analogy, to make appropriate decisions without human guidance.[4]

Computers may be modeled on human brains, but human minds do not work just like computers. We are learning, for example, that our cognitive function, self-reflection, and rational thinking depend on emotions. Stories appeal to emotions, *pathos*, the root of *empathy*, sharing feelings. Stories continue to be alive as long as they summon strong, complicated emotions, as long as they still resonate with real dilemmas, and as long as they are good to think with. We have seen how Greeks and other ancient societies told themselves stories to try to understand humankind's yearning to exceed biological limits and to imagine the consequences of those desires. The insights and wisdom in such myths might deepen our discourse about AI.

Biotechne stories, perpetuated over millennia, are a testament to the persistence of thinking and talking about what it is to be human and what it means to simulate life. We are hardwired to hear, tell, and remember stories. As George Zarkadakis reminds us, stories "are the most powerful means available to our species for sharing values and knowledge across time and space."[5] This raises an intriguing possibility.

Might myths about artificial life in all its forms, like the examples gathered in this book, play a role in teaching AI to better understand humankind's conflicted yearnings? Perhaps some day AI entities will be able to absorb mortals' most profound wishes and fears as expressed in mythic musings about artificial life. Perhaps AI beings might somehow grasp the tangled expectations and fears we have of AI creations. Through learning that humans foresaw their existence and contemplated some of the quandaries the machines and their makers might encounter, AI entities might be better able to comprehend—even "empathize" with—the quandaries that they pose for us.

The rise of a Robot–Artificial Intelligence "culture" no longer seems far-fetched. AI's human inventors and mentors are already building the Robot-AI culture's *logos* (logic), *ethos* (moral values), and *pathos* (emotions). As humans are enhanced by technology and become more like machines, robots are becoming infused with something like humanity. We are approaching what some call the new dawn of Robo-Humanity.[6] When that day comes, what myths and stories will we tell ourselves? The answers will shape how and what our AI creations will learn too.

GLOSSARY

agency. The capacity, condition, or state of acting, operating, or exerting power or energy in a given environment.

android, droid. A mobile robot in human form.

Artificial Intelligence (AI). Intelligence, or mind, displayed by artificial life or machines, analogous to the natural intelligence of animals and humans; capable of perceiving its environment and taking action. AI mimics cognitive functions associated with mind, such as learning and problem solving. "Narrow AI" allows a machine to carry out specific tasks, while "general AI" is a machine with "all-purpose algorithms" to carry out intellectual tasks that humans are capable of, with abilities to reason, plan, "think" abstractly, solve problems, and learn from experience. AI can also be classified by types: Type I machines are reactive, acting on what they have been programmed to perceive at the present, with no memory or ability to learn from past experience (examples include IBM's Deep Blue chess computer, Google's AlphaGo, and the ancient bronze robot Talos and the self-moving tripods in the *Iliad*). Type II AI machines have limited capacity to make memories and can add observations to their preprogrammed representations of the world (examples: self-driving cars, chatbots, and Hephaestus's automated bellows). Type III, as yet undeveloped, would possess theory of mind and the ability to anticipate others' expectations or desires (fictional examples: *Star Wars*' C-3PO, Hephaestus's Golden Servants, the Phaeacian ships). Type IV AI of the future would possess theory of mind as well as self-awareness (fictional examples include Tik-Tok in John Sladek's 1983 novel and Eva in the 2015 film *Ex Machina*). Since she is capable of deceit and persuasion, Pandora seems to fall between Types II and III.

artificial life. Systems, beings, or entities that simulate natural life, natural processes; or replicate aspects of biological phenomena; human or animal artifacts brought to life.

automation. The technology by which action is performed without human assistance.

automaton, automata. A self-moving mechanical or constructed device, usually resembling an animal or human, that is not directly operated by an agent. Some automata are machines that perform tasks according to predetermined instructions; some automata can respond with a range of responses to different circumstances.

bionic. Having artificial body parts that amplify human or animal powers.

biotechne. Ancient Greek *bio*, life, *techne*, craft, art, science, the application of knowledge to practice.

biotechnology. Technology based on manipulating biological organisms, living systems, or their components to develop, modify, or make products or processes.

black box. Complex entity, machine, or system whose outputs are known or observable but whose inner contents and internal workings are hidden, unknown, opaque, and mysterious to the user.

cyborg, cybernetic organism. A being, usually humanoid, that combines or integrates organic, biological components with artificial technology, a human-machine hybrid, often exceeding human capabilities.

device. An object, gadget, instrument, contrivance, or apparatus made for a particular purpose, often denoting a mechanical item.

fembot. A robot in the form of a human female.

machine. A mechanical structure or device based on one or more components (such as lever, pulley, wheel and axle, inclined plane, screw, wedge) that changes the direction or magnitude of a force.

machine learning. Computers and AI with the ability to learn independently, without being explicitly programmed.

mechanism, mechanical. Something made of parts that move or work together to perform an action; a machine or something resembling a machine.

programmed. Supplied with a predetermined set of (coded) instructions for automatic performance.

puppet, marionette, doll. An artificial model of a human or animal typically moved by hand, rods, wires, or strings.

rejuvenation. To make a living being young again, to restore youthful strength, vigor, and/or appearance.

robot, bot. Complex and ambiguous to define, but a robot usually is a machine or self-moving object with a power source that provides energy. It can be "programmed" to "sense" its surroundings, and has a kind of "intelligence" or way of processing data to "decide" to interact with the environment to perform actions or tasks. Talos, the bronze animated statue powered by ichor, fits this definition.

Uncanny Valley. The eerie and repellent sensation experienced by most human beings when encountering artificial life forms, especially humanoid entities, that appear to be almost but not precisely real. Affinity increases with verisimilitude but steeply drops off as the entity approaches being indistinguishable from reality. The hypothesis was first identified by roboticist Masahiro Mori in 1970.

NOTES

CHAPTER 1. THE ROBOT AND THE WITCH: TALOS AND MEDEA

1. Apollonius *Argonautica* 4.1635–88; Apollonius (Hunter trans.) 2015, 6, 298–304. The Greek word *automaton*, "acting of one's own will," was first used in Homer *Iliad* 5.749 and 18.371-80 to describe the automatic door opening and automatic wheeled tripods built by Hephaestus for the gods; see chapter 7. Hound and javelin, Ovid *Metamorphoses* 7.661-862.
2. On the "slippery" terms *robot* and *automaton* for an ancient "object constructed to move on its own," see glossary; cf. Bosak-Schroder (2016, 123, 130–31), who argues that the earliest automata in Greek literature were originally imagined as solely magical and only later attained mechanical life. The idea of automated tools that can finish a task without continued human input, along with the impulse to make them, is very ancient, beginning with the Stone Age *atlatl* (spear thrower) and the bow and arrow. Once the arrow is nocked, aimed, and released, the bow fires "this little spear further, straighter, and more consistently than human muscles ever could," remarks Martinho-Truswell (2018).
3. For a classicist's perspectives on Harryhausen's Talos: Winkler 2007, 462–63.
4. Hesiod *Works and Days* 143–60. In Hesiod's poem, the "Age of Bronze" was a symbolic chronology of the warlike Bronze Age generations that preceded present-day Iron Age humans; Apollonius's poetic license makes the men of that age literally of bronze. Gantz 1993, 1:153. There was also a legendary Athenian inventor named Talos; see chapter 5. Various genealogies of Talos: Buxton 2013, 77–79.
5. Ancient Colchis is now the Republic of Georgia. "Medea's oil," *Suda* s.v. Medea.
6. Apollodorus *Library* 1.9.26; Apollonius *Argonautica* 3.400–1339.
7. Medea's *technai*, devices: Pindar *Pythian* 4.
8. Another version of Medea and her relationships with Jason and the Argonauts: Diodorus Siculus 4.45–48. Motif of heroes' and monsters' sole vulnerability, Buxton 2013, 88–94.
9. Colossus of Rhodes, Pliny 34.18; Strabo 14.2.5. N. F. Rieger in Ceccerelli 2004, 69–86. Centuries earlier, Rhodes was also famous for its "living statues"; see chapters 5 and 9.
10. Why people tend to attribute life to machines and Artificial Intelligence, Bryson and Kime 2011; Shtulman 2017, 138; Zarkadakis 2015, 19–23, 25–27. Trust and empathy in human-robot interactions: Darling, Nandy, and Breazeal 2015; Lin, Abney, and Bekey 2014, 25–26; and Lin, Jenkins, and Abney 2017, chapters 7–12.

When "thinking machines express anxiety about their own demises" it is "surely a sign of 'consciousness' "; Mendelsohn 2015. Can Artificial Intelligence be tricked? Reynolds 2017.

11. Sophocles *Daedalus* fr. 160, 161 R. Winkler 2007, 463.
12. In a story mentioned by Apollodorus (*Library* 1.9.26), the Argonaut Poeas shot Talos in the ankle, which recalls the death of the mythic hero Achilles by a poison arrow to his vulnerable heel. Rock-throwing giants were a common motif in ancient myth and art. Another source says Talos was a bronze bull, perhaps conflating him with the Minotaur, the bull-headed man kept by Minos in the Cretan Labyrinth (see chapter 4). Coins of Knossos show the Minotaur throwing stones, and some Talos coins of Phaistos show a bull on the reverse.
13. Ganz 1993, 1:365. Robertson 1977. Teardrop: Buxton 2013, 82 and fig. 3 caption. Metallic objects and statues were often painted whitish in red-figure vase iconography; for example, several images of Niobe being turned into stone show her body partly white. Another notable detail is the ornamental border around the top of the Ruvo krater that appears to represent blacksmith's tongs; see figs. 7.4 and 7.5, and the similar design in the border at the top of the Niobe Painter's krater depicting Pandora, who was also fabricated by Hephaestus, fig. 8.7.
14. Robertson 1977, 158–59. Buxton 2013, 81 and figs. 4–6.
15. Carpino 2003, 35–41, 87, quote 41. Medea and local Etruscan versions of Greek myths, de Grummond 2006, 4–5.
16. Gantz 1993, 1:341–65, on artistic and literary sources for Talos; Apollonius *Argonautica* 4.1638–88; Simonides fr. 568 PMG; Apollodorus *Library* 1.9.26 and J. Frazer's note 1; 1.140; Photius *Bibliotheca* ed. Bekker, p. 443b, lines 22–25; Zenobius *Cent.* v. 85; Eustathius scholiast on *Odyssey* 20.302. Divine robotic devices are discussed in chapter 7.
17. Faraone 1992, 41. Quotes, Hallager 1985, 14, 16–21, 22–25. Cline 2010, 325, 523. For photos and a drawing of the Master Impression seal, Chania Museum of Archaeology, Crete, see CMS VS1A 142 at Arachne.uni-koeln.de.
18. Shapiro 1994, 94–98, on the lost *Argonautica* epic cycle.
19. Simonides fr. 204 PMG; scholion to Plato *Rep.* 337a. Blakely 2006, 223. Sardinia and Crete, Morris 1992, 203. Etruscans and Nuragic Sardinian links: http://www.ansamed.info/ansamed/en/news/sections/culture/2018/01/08/etruscan-settlement-found-in-sardinia-for-first-time_288c45c9-9ae3-4b5e-ab8d-cb9bf654b775.html.
20. Laestrygonians are also described by Apollodorus *Epitome* 7.13; Thucydides 6.2.1; Hyginus *Fabulae* 125; Ovid *Metamorphoses* 14.233; Strabo 1.2.9. A pair of wall paintings, ca. 50–40 BC (Vatican Museum, Rome), depicts the Laestrygonians as copper-colored giants wresting up boulders and heaving them at Odysseus's sailors. Paratico 2014.
21. Kang 2011, 15–16, 19, 21, 312nn1–3.
22. Weinryb 2016, 154.
23. Gods don't use technology; Talos is "biological" and not an automaton because an automaton must have "an internal mechanism," Berryman 2003, 352–53; Aristotle

on automaton "self-moving" puppets, 358. Devices made by Hephaestus are "animated by divine power," not technology, and gods do not use technology, Berryman 2009, 25–26 (Talos is omitted from discussion). Cf. Kang 2011, 6–7 and 311n7. But see De Groot 2008 and Morris 1992 on the overwhelming evidence from ancient literature and art that Greek gods were imagined as using technology and tools in projects, including self-moving entities. "Mechanistic" analogies could arise before "full-fledged automata" were feasible.

24. Bosak-Schroeder 2016, 123, 132. Cf. Berryman 2009, 22, "mechanistic conceptions" could not have been imagined before mechanics developed "as a discipline." Contrast Martinho-Truswell 2018 on prehistoric inventions and see Francis 2009; archery, catapults, voting machines, and the winepress demonstrate practical mechanics.
25. Definition, Truitt 2015a, 2. Ancient Greek automata as "self-moving," Aristotle *Movement of Animals* 701b.
26. This quote is from Berryman 2007, 36; Aristotle on natural and unnatural life, 36–39.
27. Truitt 2015b, commenting on Cohen 1963.
28. The myths of Pandora, Talos, the Golden Maidens, and other androids "distinguish these simulations, these artificial 'humans' from organic, natural life forms by the composition of the body," not necessarily by "mechanistic" qualities. "Artificial life, in these myths, is made of the same substances" and methods "that human craftsmen use to make tools, buildings, and artworks" and statues. As with robots today, their functions are "labor, defense, and sex." Raphael 2015, 186. See Berryman 2009, 49 and n119, *techne* is better translated as science rather than art.
29. Popular links between metalworking and magic are widespread: Blakely 2006; Truitt 2015b; Truitt 2015a, guarding borders, 62–63; Faraone 1992, 19 and 29n11, 18–35. Weinryb 2016, 109, 128–34.
30. Blakely 2006, 81, 209. Weinryb 2016, 153, 53–54, 154–56. Clarke 1973, 14, 21, 36.
31. On the history of ancient Greek belief in the agency of statues, Bremmer 2013.
32. Blakely 2006, 210–12.
33. Cook 1914, 1:723–24; Buxton 2013, 86–87; Weinryb 2016, 4–7, 14, 44–52.
34. Lost-wax process: Mattusch 1975; Hodges 1970, 127–29. Bronze techniques using wax and clay models, Hemingway and Hemingway 2003. Wooden armatures, see chapter 6. Realistic bronze statues from plaster casts of humans, chapter 5 and Konstam and Hoffmann 2004.
35. Raphael 2015, 187. Berryman 2009, 27. Mayor 2007; Mayor 2016.
36. Apollonius (Hunter trans.) 2015, 300; Raphael 2015, 183–84;. Aristotle on automata, puppets, biology, physiology, and mechanics, Leroi 2014, 172–73, 199–202. De Groot 2008.
37. Ichor: Homer *Iliad* 5.364–82. "Talos in fact has ichor, rather than blood in his vein," although we "should perhaps not enquire too closely as to what flowed in Talos's vein," notes R. Hunter trans., *Apollonius* 2015, 189, 300, 304. Ichor in myth and medical treatises, Buxton 2013, 94–96.

38. Bloodletting was thought to have beneficial value in healing various ailments. Hippocrates *On the Nature of Man* 11; Aristotle *History of Animals* 512b 12–26. Bloodletting is depicted on the Peytal Aryballos, 480 BC, Louvre. Buxton 2013, 93. The location of Talos's weak point, the ankle, conforms to the trope of vulnerability associated with feet, e.g., Achilles's heel and Oedipus's lame foot.
39. Plutarch *Moralia* 5.7.680C–83B; Dickie 1990 and 1991; Apollonius (Hunter trans.) 2015, 6, 302. On bronze and evil eye, Weinryb 2016, 131–33. Examples of realistic painted and inlaid bronze statues, Brinkmann and Wünsche 2007.
40. Truitt 2015a and b. Kang 2011, 22–25, 65–66. Buxton 2013, 74. Gray 2015. "In-betweenness" of Pandora, chapter 8 and Francis 2009, 14–15. In a sense, Talos could be said to have "narrow" or Type I reactive AI (see glossary). On the "Uncanny Valley" effect of realistic artificial life, see chapter 5; and Lin, Abney, and Bekey 2014, 25–26.
41. Newman 2014. The myth of Talos as an invincible ancient security system underlies the name of the "world's largest hub of security intelligence" working "tirelessly to identify and counter cyber-crime attacks," called Talos, maintained by Cisco Systems, since 2008. http://www.talosintelligence.com/about/.
42. Kang 2011, 65. On modern concerns about the ethics of replacing human judges with AI, see Bhorat 2017. Lin 2015; Lin, Abney, and Bekey 2014, 53, 60, and chapters 4 and 5. Thanks to Norton Wise for valuable suggestions on these questions. Spenser's Iron Knight, Talus, was named for the mythic Talos but may have been modeled in part on Leonardo da Vinci's robotic knight in armor (ca. 1495) clad in heavy medieval armor and powered by pulleys, cranks, gears, and levers.
43. See chapter 9 for ancient Persian "batteries." Ambrosino 2017. Shtulman 2017, 53–56.
44. Tenn 1958. Talos served as "a primitive home alarm system," Mendelsohn 2015.
45. Garten and Dean 1982, 118. Talos missiles were decommissioned in 1980. Talos in the Harryhausen film of 1963 also combined preprogrammed "brawn" with "brains." Winkler 2007, 462–63.
46. History of efforts to create military robotics, Jacobsen 2015 and Tyagi 2018. Nissenbaum 2014. SOCOM TALOS project renewed its official call for proposals in December 2017–18.

CHAPTER 2. MEDEA'S CAULDRON OF REJUVENATION

1. Ovid *Metamorphoses* 7.159–293.
2. *Nostoi* frag. 7, and Medea's plot against Pelias in the lost play by Sophocles, *Rhizotomoi*, "Root-Cutters," see Gantz 1993, 1:191, 367; some accounts indicate that she placed Aeson in the boiling kettle. Godwin 1876, 41.
3. Medea's rejuvenation plan in the Aeschylus play, according to scholia on Euripides *Medea*, see Denys Page, ed., *Euripides, Medea* (Oxford, 1938). Diodorus Siculus 4.78 on the revivifying effects of the steam bath invented by Daedalus. New technologies often misconstrued, Hawes 2014, 59–60; on Palaephatus and his date, see 37–91 and 227–38. Aristotle on metabolism, aging, and life spans, Leroi 2014, 260–65.

4. Ovid *Metamorphoses* 7.159–293; Clauss and Johnston 1997, 33–34; Godwin 1876, 41; Newlands 1997, 186–92. Only mercury corrupts gold. Maluf 1954. Exchange transfusions are lifesaving procedures for sickle-cell anemia and blood diseases of newborns. Blood exchange parabiosis experiments, in which young blood is transfused into an older body, Friend 2017, 60–61. Older mouse tissues were rejuvenated but the young donor mice aged faster.
5. Psamtik's suicide by drinking bull's blood, Herodotus 3.15.4; Plutarch *Themistocles* 31; and Midas, see Strabo 1.3.21. Stormorken 1957.
6. See "Ruse of the Talismanic Statue," Faraone 1992, 100–104.
7. Faraone 1992, 100.
8. Quotes from Diodorus Siculus 4.50–52; other sources include Pindar *Pythian* 4.138–67; 4.249–50; Apollonius of Rhodes *Argonautica* 4.241–43; Apollodorus *Library* 1.9.27–28; Ovid *Metamorphoses* 7.159–351; Pausanias 8.11.2–3; Hyginus *Fabulae* 21–24. A lost play of 455 BC by Euripides, *Peliades*, dramatized this myth. Gantz 1993, 1:365–68. Medea's transformation mirrors the goddesses' use of ambrosia as a rejuvenating salve, Homer *Iliad* 14.170 and *Odyssey*18.188.
9. Diodorus Siculus (4.52.2) suggests that Medea hypnotized the daughters and created the illusion (*eidolon*) of a young lamb emerging from the pot.
10. Examples include an Etruscan olpe, Oriental style, ca. 630 BC with incised image of Medea inscribed "Metaia," black *bucchero*, from Caere (Cerveteri), Museo Archeologico Nazionale inv. 110976; de Grummond 2006, 4–6 and fig. 1.7. Two black-figure vases from Vulci show Medea and a ram in the cauldron in the British Museum, B 221 and B 328; black-figure vase has similar images by the Leagros Group, in the Harvard University Art Museum, 1960.315.
11. Red-figure krater in Boston Museum of Fine Arts, 1970.567; red-figure vase from Vulci, ca. 470 BC, British Museum E 163. Woodford 2003, 80–83, fig. 54, red-figure cup, 440 BC, Vatican Museum.
12. Dolly was cloned from an adult cell (cows had previously been cloned) by the Roslin Institute, University of Edinburgh. Dolly and other cloned sheep in the project died of a fatal contagious virus, but a 2016 study by Sinclair et al. of Dolly's skeletal remains (stored in the National Museum of Scotland) did not reveal evidence of premature aging of her bones. http://www.roslin.ed.ac.uk/public-interest/dolly-the-sheep/a-life-of-dolly/.
13. Buddhist perspectives on replicating life and cloning, see Han 2017, 67.
14. Apollodorus *Epitome* 5.5; scholiast on Apollonius *Argonautica* 4.815. Medea contemplates suicide in *Argonautica* 3.800–815.
15. On promotions of mortals to immortality, Hansen 2004, 271–73. Iolaus: Pindar *Pythian* 9.137; Euripides *Heraclidae*.
16. Ovid *Metamorphoses* 7.171–78; Newlands 1997, 186–87. In Homer's *Odyssey* 7.259, the witch-nymph Calypso's offer of immortality to Odysseus was seen as "irrational" by the skeptic Heraclitus: Hawes 2014, 96. See chapter 3 for that story.
17. Chiron, Apollodorus *Library* 2.5.4.
18. Dioscuri, Apollodorus *Library* 3.11.2.

CHAPTER 3. THE QUEST FOR IMMORTALITY AND ETERNAL YOUTH

1. Mayor 2016. "Cheating Death" 2016. Raphael 2015, 192–93. Boissoneault 2017. *Blade Runner* was loosely adapted from the science-fiction novel *Do Androids Dream of Electric Sheep?* by Philip K. Dick (1968). In Jo Walton's science-fiction novel set in antiquity, *The Just City* (2015), 254, 300, robot-slaves are punished by having their memories deleted. In the popular TV series *Westworld* (HBO, 2016 premiere) the androids' memories are swept clean each day.
2. Lefkowitz 2003, 90–91. Reeve 2017. Rogers and Stevens 2015, 221–22.
3. Aristotle (*On the Soul* 2.2.413a21–25) defines a living thing as able to take in nutrition (lowest common denominator) and to change (plants), capable of movement, motivation or desire, and perception (animals), and, for humans, having the added capacity for thought. For Aristotle, plants and animals change, but artificial artifacts cannot change. Steiner 2001, 95. Exceptions include Hephaestus, who is lame and hardworking; see chapter 7.
4. The Titan Prometheus is an exception—his aid to humans entailed high-stakes risks, and his immortality would be part of the punishment. John Gray's *Soul of the Marionette* (2015) explores human freedom and immortality through the lens of Gnosticism.
5. Cave 2012, 6–7, 202, 205–9. Gilgamesh and immortality, Eliade 1967. Amazons die as heroes, Mayor 2014, 28–29.
6. Colarusso 2016, 11.
7. Hansen 2002, 387–89. Human life span of 120 years, Zimmer 2016.
8. Pindar cited by Pausanias 9.22.7; Plato *Republic* 611d; Ovid *Metamorphoses* 13.904–65. Palaephatus 27 *Glaukos of the Sea*. Glaukos, Hyginus *Fabulae* 136; Apollodorus *Library* 3.3.1–2.
9. *Alexander Romance* traditions, Stoneman 2008, 94, 98–100, 146–47; 150–69. Aerts 2014, 498, 521.
10. In the *Classic of Mountain and Seas*, Birrell 1999, 241.
11. Mercury fumes can be lethal but ingestion is not. Qin Shi Huang: Kaplan 2015, 53–59; Cooper 1990, 13–28; 44–45.
12. Alexander quotes Homer *Iliad* 5.340. The story appears in Plutarch *Moralia* 341b, *Moralia* 180e, and Plutarch *Alexander* 28, among others. Buxton 2013, 95–96.
13. Homer *Odyssey* 24.5.
14. Stoneman 2008, 152–53.
15. Gantz 1993, 1:154–56. Apollodorus *Library* 1.7, 2.5.4. Hard 2004, 271. Kaplan 2015, 24–28. Simons 1992, 27. Hyginus (*Astronomica* 2.15) says the torment lasted 30,000 years, elsewhere 30 years. Strabo (11.5.5) says 1,000 years. Liver regeneration is reflected in Chinese folklore in the utopian figure of *shih-jou*, a mound of meat that looks like ox liver and can never be completely consumed because it regenerates, Birrell 1999, 237.
16. Heracles and the Hydra, Hard 2004, 258. Mayor 2009, 41–49.

17. Hansen 2002, 36–38. Felton 2001, 83–84.
18. Sisyphus: Apollodorus *Library* 1.9.3–5 and Frazer's note 3, Loeb ed., pp. 78–79; Homer *Odyssey* 11.593–600; scholiasts on Homer *Iliad* 1.180 and 6.153; Pherecydes *FGrH* 3 F 119.
19. *Homeric Hymn to Aphrodite* 218–38; Apollodorus *Library* 3.12.4 and Frazer's note 4, Loeb ed., pp. 43–44. In antiquity, cicadas were associated with renewed youth and living forever, sloughing off old skin and emerging anew. Tithonus and Eos in classical art and literature, Gantz 1993, 1:36–37. Woodford 2003, 60–61. Lefkowitz 2003, 38–39.
20. Hansen 2004, 222, 273. Cohen 1966, 15, 16, 24.
21. Hansen 2004, 269–73. *Homeric Hymn to Aphrodite* 239–48.
22. Eos and Tithonus in medieval and modern arts, Reid 1993, 1:386–88.
23. Sappho's Tithonus poem, West 2005, 1–9. D'Angour (2003) discusses Horace's ode in view of Pythagorean notions. Tennyson's "Tithonus," Wilson 2004, 214n78. Ageless longevity is a universal theme in the folklore of utopias, Stoneman 2008, 99–100; 153–54. De Grey 2008 and 2007. In the final novel of Philip Pullman's *His Dark Materials* trilogy (1995, 1997, 2000), God himself is revealed as a "twittering ghost."
24. Leroi 2014, 260–65. Friend 2017, link between sexual abstinence and extending life, 65. Named for the mythic afterlife of heroes, "Elysium" health supplements aim to guarantee "overliving": https://www.fastcompany.com/3041800/one-of-the-worlds-top-aging-researchers-has-a-pill-to-keep-you-feeling-young.
25. "Life detested," Woodford 2003, 60. On anxiety ancient and modern about technoculture's threat to "human finitude" and "humanity," Cusack 2008, 232.
26. Cave 2012. Friend 2017. Harari 2017, 21–43. Buddhist transhumanism, Mori 2012; Borody 2013. What is the limit for human longevity? Scientists debate this controversial question; some findings suggest that the maximum life span with current technology is about 115–20 years: Zimmer 2016.
27. "The disposable soma" springs the "trap of Tithonus": "Cheating Death" 2016 and "Longevity" 2016. Liu 2011, 242–43. Richardson 2013. Kaplan 2015, 68–73. Cave 2012, 64, 67–71. Friend 2017, 56–57; de Grey 2007, 8 and 379n2; de Grey 2008, "global nursing home."
28. The replicants of *Blade Runner* die too soon, before they can become human, Raphael 2015. Talos, Buxton 2013, 78. The ancient Greek concept of living too long is explored through the mythic figures of Oedipus and Heracles and Shakespeare's Macbeth and Lear in Wilson 2004, 2, 207nn2–3, 214.

CHAPTER 4. BEYOND NATURE: ENHANCED POWERS BORROWED FROM GODS AND ANIMALS

1. Plato's legend and pre-Socratic writings, Gantz 1993, 1:166. Plato *Protagoras* 320d–321e. The etymologies are Plato's, accepted in antiquity. In some ancient traditions, it was Prometheus who made the first humans and animals; see chapter 6 and Tassarini 1992, 61–62, 78–80.

2. Rogers and Stevens 2015, 1–3. On modern "Human Enhancement Technologies [HET]," see Lin 2012 and 2015. Martinho-Truswell 2018 points out that many creatures use tools, but humans are the only animals who "automate" tools, and the impulse is at least as old as the first *atlatl* and bow and arrow.
3. Prosthetics in ancient myth and history: James and Thorpe 1994, 36–37: LaGrandeur 2013. Zarkadakis 2015, 79–82.
4. Lin 2012; Patrick Lin is director of the Ethics + Emerging Sciences Group, California Polytechnic State University. History of religious qualms about artificial human enhancements and robots: Simons 1992, 28–32.
5. Ancient technology, Brunschwig and Lloyd 2000, 486–94.
6. Gantz 1993, 1:359–63. Medea collecting the Promethean drug from the gore of his liver was taken up by later authors: Propertius *Elegies* 1.12; Seneca *Medea* 705; Valerius Flaccus *Argonautica* 7.352. The ichor of the primeval giants killed by the gods spilled into the ground, causing evil-smelling springs, a belief reported by Strabo 6.3.5.
7. Apollonius, *Argonautica* 3.835–69; 3.1026–45; 3.1246–83. Pindar, *Pythian* 4.220–42. The tasks set for Jason by Aeetes were dramatized by Sophocles in his lost play *Colchides* ("The Colchians"), probably the source for Apollonius, Gantz 1993, 1:358–61.
8. Zarkadakis 2015, 79–82. Harari 2017, 289–91. See Lin 2012, 2015; for a series of reports and articles on the grave ethical issues surrounding "supersoldiers" and cyber weapons and enhancing fighters through technology and drugs, see Ethics + Emerging Sciences Group, http://ethics.calpoly.edu/he.htm. Research on neurocomputer technology to delete thoughts threatens mental integrity and cognitive liberty, Ienca and Andorno 2017.
9. The fire-breathing bulls episode also appears in Pindar *Pythian* 4.224–50 (ca. 462 BC), Shapiro 1994, 94–96.
10. Apollonius *Argonautica* 3.401–21; 3.492–535; 3.1035–62; 3.1170–1407. Godwin 1876, 41. This tactic is the same one that saved the hero Cadmus in Thebes. In that myth, Cadmus casts rocks among the Spartoi, "Sown Men," who spring up from the planted teeth of another slain dragon. Rationalizing of the sown men, Hawes 2014, 140–41, 146.
11. Mayor 2016.
12. Mayor 2009, 193–94; Stoneman 2008, 77; Aerts 2014, 255.
13. Mayor 2009, 235–36, fig. 39, illustration of Alexander's fire-breathing iron riders and horses on wheels in Firdowsi's *Shahnama* manuscript of Great Il-Khanid AD 1330–40, Sackler Museum, Harvard University.
14. It is interesting that Firdowsi's epic also describes an enchanted castle defended by automaton-archers. A later sixteenth-century illustrated manuscript shows the automatic archer shooting arrows at an invading army from its post on the castle walls; *Shahnama* by Firdowsi, Moghul, sixteenth-century illustrated MS 607, fol. 12v, Musée Condé, Chantilly, France.
15. Cusack 2008, on Talos, Nuada, Freyja, and the Hindu Savitr.
16. *Rig Veda* 1.13, 1.116–18, 10.39. Prosthetics technologies, Zarkadakis 2015, 79–81.

17. These and the following archaeological examples of prosthetics, see Nostrand 2015.
18. James and Thorpe 1994, 36–37. Egyptian toe, Voon 2017. Nostrand 2015. Mori 2012; Borody 2013.
19. Cohen 1966, 16–18. Morris 1992, 17–35, 244–50; Hawes 2014, 49–53, 207–12; "first inventor motif," 59–60, 109, 120–21, 210–11, 230–31. First "hero" inventor, Kris and Kurz 1981; "archetypal craftsman," Berryman 2009, 26. Lane Fox 2009, 186–91. Ancient sources for Daedalus's works, Pollitt 1990, 13–15. In the *Classic of Mountain and Seas*, Chinese mythology designates several inventor gods and culture heroes, such as Hsien-yuan, "Cart Shaft," who first harnessed animals to draw vehicles; Chi Kuang, "Lucky Glare," inventor of the chariot; Chi'iao Ch'ui, "Skill Weights," god of inventive technology, Birrell 1999, 205, 220, 239, 256.
20. Apollodorus *Library* 3.15.1; Antoninus Liberalis *Transformations* 41.
21. *Spy in the Wild*, BBC-PBS Nature miniseries, 2017, features more than thirty animatronic creatures fitted with cameras to secretly observe animals in nature; the animals accept and interact with the robots, even mourning their "death." Artistic works that deceive humans and animals in antiquity, Morris 1992, 232, 246. Spivey 1995.
22. Pornography and automata, Kang 2011, 108, 138–39, 165–66; Lin, Abney, and Bekey 2014, 58, 223–248; Higley 1997. Morris 1992, 246 on erotic interaction with lifelike statues; cf. Hersey 2009 and Wood 2002, 138–39.
23. Sources for the myth include Palaephatus 2 and 12; Apollodorus *Library* 3.1.3–4; Hyginus *Fabulae* 40; Hesiod frag. 145 MW; Bacchylides 26; Euripides's lost play *The Cretans*; Sophocles's lost play *Minos*; Isocrates 10 *Helen* 27; Diodorus Siculus 4.77; Ovid *Metamorphoses* 8.131–33 and 9.736–40; Ovid *Ars Amatoria* 1.289–326.
24. "Relief skyphos with Pasiphae, Daedalus, and the Heifer," Los Angeles Museum of Art, AC1992.152.15; Roman mosaic floors, House of Poseidon, second century AD, Zeugma Mosaic Museum, Gaziantep, Turkey; third century AD, Lugo, Spain; Roman frescoes, first century AD, in Herculaneum and in Pompeii's House of the Vettii (which shows the bow-drill) and Casa della Caccia Antica. De Puma 2013, 280. Pasiphae in medieval and modern arts, Reid 1993, 2:842–44.
25. Pasiphae and the Minotaur in ancient literature and art, Gantz 1993, 1:260–61, 265–66. Woodford 2003, 137–39. Rationalization in antiquity, Hawes 2014, 58, 126–27. Other ancient instances of humans copulating with animals such as horses and donkeys were reported, e.g., in Plutarch's *Moralia, Parallel Stories* 29.
26. Gantz 1993, 1:261–64, 273–75.
27. Ancient Scandinavian sagas tell of the blacksmith Wayland who devised wonderful weapons and other marvels, including a garment made of real birds' feathered skins, which allowed him to fly, Cohen 1966, 18.
28. Daedalus and Icarus ancient sources and art, Gantz 1993, 1:274–75; in medieval and modern arts, Reid 1993, 1:586–93. Beeswax and feathers were said to be the building materials of one of the first temples to Apollo, according to Pindar and other poets, Marconi 2009.
29. Morris 1992, 193.

30. Etruscan *bucchero* olpe found at Cerveteri (Caere), ancient Etruria, Lane Fox 2009, 189. Boeotian Corinthianizing alabastron of ca. 570 BC, in Bonn. Etruscan bulla, Walters Art Museum, Baltimore, 57.371. Morris 1992, 194–96. Daedalus on Etruscan gems, Ambrosini 2014, 176–78, and figs. 1–15b.
31. Icarus and Daedalus in art, Gantz 1993, 1:274; *LIMC* 3. "Fall of Icarus," seascape wall painting from Pompeii, National Archaeological Museum of Naples. On the widespread folklore motif of an architect devising a way to fly from captivity, see Kris and Kurz 1979, 87–88.
32. Flying in Greek comedy: D'Angour 1999. Keen 2015, 106–19.
33. Stoneman 2008, 111–14. Aerts 2014, 27.
34. Stoneman 2008, 114–19. For medieval images of Alexander as aviator, Schmidt 1995.
35. Needham and Wang 1965, 587–88.
36. *Classic of Mountain and Seas*, Birrell 1999, 256.
37. Recorded in *Zizhi Tongjian*, the historical chronicle of Chinese history 403 BC to AD 959, compiled in AD 1084. Other ancient myths of flight by men, Cohen 1966, 95–96. See chapter 9 for forced flying punishments of criminals.
38. Among the ancient texts that discuss Daedalus's flight are Apollodorus *Epitome* 1.12–15; Strabo 14.1.19; Lucian *Gallus* 23; Arrian *Anabasis* 7.20.5; Diodorus Siculus 4.77; Ovid *Metamorphoses* 8.183, *Heroides* 4, *Ars Amatoria* 2, *Tristia* 3.4; Hyginus *Fabulae* 40, Virgil *Aeneid* 6.14. McFadden 1988.

CHAPTER 5. DAEDALUS AND THE LIVING STATUES

1. Daedalus and Sardinia, Morris 1992, 202–3, 207–9; Diodorus Siculus 4.30; Pausanias 10.17.4. Tools, Vulpio 2012. The Nuragic iron compass is in Sanna Museum, Sassari, Sardinia.
2. Diodorus Siculus 4.78. See Morris 1992 for all the inventions attributed to Daedalus.
3. Blakemore 1980.
4. Michaelis 1992. Ayrton 1967, 179–84. Ayrton's controversial modernist sculpture of the bronze robot Talos stands guard on Guildhall Street, Cambridge, UK.
5. Honeycomb building blocks, Marconi 2009. Marcus Terentius Varro's conjecture, in *On Agriculture*, was proven by Hales 2001.
6. Lane Fox 2009, 190.
7. The shell and ant: Zenobius *Cent.* 4.92; also mentioned in Sophocles's lost play *The Camicians*, Athenaeus 3.32.
8. For Daedalus's time in Sicily, Morris 1992, 193–210. Apollodorus *Epitome* 1.14–15; Herodotus 7.169–70. Diodorus Siculus 4.78–79 gives a slightly different version of the events.
9. Apollodorus *Library* 3.15.8; Diodorus Siculus 1.97, 4.76–77; Pliny 36.9; Pausanias 1.21.4; Ovid *Metamorphoses* 8.236; Plutarch *Theseus* 19. This Athenian Talos is sometimes called Kalos or Perdix. Some versions say the saw was modeled on a fish spine. Daedalus in Athens, Morris 1992, 215–37; folding chair, 249–50; Talos grave, 260. There is no ancient account of the death of Daedalus.

10. Pseudo-Aristotle *On Marvelous Things Heard* 81; Stephanus of Byzantium s.v. Daedalus; Diodorus Siculus 1.97; Scylax *Periplus*; Pausanias 2.4.5 and 9.40.3. Daedalus statues, Donohue 1988, 179–83.
11. Bremmer 2013, 10–11. Several ancient accounts tell of statues of gods that were bound or fettered. Lucian *Philopseudes* (second century AD) satirizes beliefs in animated statues that arise at night to bathe, sing, wander, and foil thieves; Felton 2001. Vase paintings of animated statues coming to life on buildings, Marconi 2009.
12. Morris 1992, 30–31, 221–25, 360.
13. Socrates on Daedalus, Morris 1992, 234–37; 258–89 for the Attic deme Daedalidae; Daedalus in Athens, 257–68. Kang 2011, 19–21, Socrates's statement shows that automata were viewed as slaves in antiquity. Cf. Walton 2015, a science-fiction novel set in a "utopia" based on Plato's *Republic*, in which Socrates discovers that the robot-slaves, used as tools, turn out to have consciousness and a desire for freedom.
14. Bryson 2010; Lin 2015; "AI in Society: The Unexamined Mind" 2018.
15. Semen as the liquid that animates an embryo, Leroi 2014, 199. Quote, Berryman 2009, 72.
16. Keyser and Irby-Massie 2008, s.v. Demokritos of Abdera, 235–36. Kris and Kurz 1979, 67–68. Leroi 2014, 79–80, 199–200; Kang 2011, 19–20 (erroneously claims that Aristotle attributed statues' movement to mercury), 98, 117–18. Berryman 2009, 26, 37, 75; noting that Aristotle uses the mercury analogy to criticize atomist theory. Morris 1992, 224–25, 232–33; Donohue 1988, 165–66, 179–83; Steiner 2001, 118–19. Semen, Hersey 2009, 69–71, 100. Democritus also studied magnets, Blakely 2006, 141 and n24.
17. James and Thorpe 1994, 131. Ali 2016, 473.
18. Blakely 2006, 16, 25, 159, 215–26.
19. Bremmer (2013) traces the chronological history and ancient sources for statues "with agency," 13–15 on sweating, weeping, and bleeding statues. See also Poulsen 1945, 182–84; Donohue 1988; Cohen 1966, 26 n26; Felton 2001; Van Wees 2013.
20. For contradictions in the artistic arguments, see Morris 1992, 240–56. Felton 2001, 79–80.
21. Berryman 2009, 27–28, original italics; it seems "very unlikely" that "mechanistic conceptions" could have developed "prior to the existence of mechanics as a discipline," 22. Some real devices invented before the time of Aristotle, such as catapults, voting machines, and wine and olive presses, could have inspired machine analogies. Cf. Francis 2009, 6–7.
22. On ancient Greeks' innovation and imagination, D'Angour 2011, 139–42. Rogers and Stevens 2015. "At the origin of any creation or invention lie the imagination and the ability to dream," notes Forte 1988, 50; inventions require the "effort of imagination."
23. Simons 1992, 40. Francis 2009. "Where science fiction leads," paraphrasing "The Next Frontier: When Thoughts Control Machines" 2018, 11.
24. On aesthetic and philosophical reactions to statues in antiquity, Steiner 2001. On various Greek artists and sculptors of lifelike artworks, see entries in Pollitt 1990. Realistic statues, Spivey 1995.

25. Haynes 2018. Pliny's artistic descriptions, books 34–36.
26. Quintilian *Inst.* 12.7–9; Lucian *Philopseudes* 18–20; Felton 2001, 78 and n10.
27. These examples and many more, in Pliny 34.19.59–35.36.71–96; painted marble, e.g., 35.40.133; the invention of ceramic portraits from shadow profiles of the living, 35.43.151. On artistic *phantasias*, Pollitt 1990, 222 and n2.
28. Plaster casts and clay and wax models of living people, Pliny 35.2.6, 35.43.151, and 35.44.153 (incorrectly cited as Pliny 36.44.153 by Konstam and Hoffmann 2004). Parrhasius, Seneca *Controversies* 10.5. Cf. earlier discussion of the "virtuosity" of the Riace sculptor, Steiner 2001. Kris and Kurz 1979.
29. Blakely 2006, 141–44, 157. Magnetic lodestone's properties were known to Thales of Miletus (sixth century BC); magnetism was described in Chinese chronicles, such as *Guiguzi* (fourth century BC) and *Lushi Chunqiu* (second century BC).
30. Lowe 2016, 249, 267. Heron of Alexandria devised a continuously hovering hollow sphere over a funnel opening of a closed vessel of boiling water, but the design is nonfeasible for a large statue; James and Thorpe 1994, 134; re-created by Kotsanas 2014, 61. Today, magnetic suspension or levitation (for example, maglev trains) can be achieved only by extremely powerful electromagnetic technologies and with rotation (as with Levitron toys).
31. Lowe 2016. Examples of floating statues, Rufinus, *Historia Ecclesiastica* ca. AD 550; Cedrenus, the Byzantine historian, ca. AD 1050, in *Synopsis Historion*; Nicephorus Callistus *Church History* 15.8. Stoneman 2008, 119, 261n38.
32. Claudian, "*De Magnete*/Lodestone," *Minor Poems* 29.22–51. Lowe 2016, 248n6.
33. The Uncanny Valley effect was first articulated by the Japanese robotics engineer Masahiro Mori in 1970, inspired by attempts to make hyperrealistic prosthetics; Mori 1981 and 2012; Borody 2013; and see also Zarkadakis 2015, 68–73; Kang 2011, 22–24, 34–35, 41–43, 47–55, 207–20; Lin, Abney, and Bekey 2014, 25–26. Wonder, *thauma*, and wondrous works, *thaumata*, especially in ancient Greek art, D'Angour 2011, 150–56. On the strong mixed emotions aroused by hyperreal, seemingly animated sculptures in classical antiquity, Marconi 2009. Liu 2011, 201–48. Wonder in Indian automata tales, Ali 2016.
34. Cohen 2002, 65–66. Cf. Mori 1981 and 2012; Borody 2013, and see also Raghavan 1952. See Liu 2011, 243–46, for discussion of the remarkably similar Chinese tale in the *Book of Liezi*.
35. Pollitt 1990, 17; 15–18 for artificial life described in Homer.
36. O'Sullivan 2000. Aeschylus *Theoroi*; Euripides *Eurystheus*; Bremmer 2013, 10–11; Marconi 2009; Morris 1992, 217–37. Faraone 1992, 37–38. Kris and Kurz 1979, 66–67. The "shock of the new" in ancient art, D'Angour 2011, 150–56.

CHAPTER 6. PYGMALION'S LIVING DOLL AND PROMETHEUS'S FIRST HUMANS

1. Hesiod *Theogony* 507–616; *Works and Days* 42–105. The final play is lost; Prometheus in ancient literature and art, see Gantz 1993, 1:152–66; Glaser and Rossbach 2011; Prometheus in modern arts, Reid 1993, 2:923–37.

2. Hard 2004, 96. Raggio 1958, 45. Sappho frag. 207 (Servius on Virgil).
3. Simons 1992, quote 28; from mud metaphor to mechanical engineering metaphors, Zarkadakis 2015, 29–34.
4. According to Aesop *Fables* 516, "The clay that Prometheus used was not mixed with water but with tears." Other sources for Prometheus's creation of humans include Menander and Philemon, per Raggio 1958, 46; Aristophanes *Birds* 686; Aesop *Fables* 515 and 530; Apollodorus *Library* 1.7.1; Callimachus frag. 1, 8, and 493; Aelian *On Animals* 1.53; Pausanias 10.4.4; Ovid *Metamorphoses* 1.82 and 1.363 (Deucalion's Flood); Horace *Odes* 1.16.13–16; Propertius *Elegies* 3.5; Statius *Thebaid* 8.295; Juvenal *Sat.* 14.35; Lucian *Dialogi deorum* 1.1; Hyginus *Fabulae* 142; Oppian *Halieutica* 5.4; Suidias (Suda) s.v. Gigantiai. Enlivened by fire: Raggio 1958, 49; Dougherty 2006, 50, citing Servius commentary on Virgil *Eclogues* 6.42.
5. Early European travelers visited the ravine: in the eighteenth century Sir William Gell reported that some stones there emitted an odor; in the nineteenth century Colonel Leake found the pair of boulders but discerned no smell; George Frazer noticed reddish earth but no large rocks. See Peter Levi's note 19 in vol. 1 of 1979 Penguin edition of Pausanias.
6. Pygmalion myth and ancient statue lust, Hansen 2017, 171–75.
7. Buddhist tale of a mechanical girl for sex, Lane 1947, 41–42, and Kris and Kurz 1979, 69–70. Ambrosino 2017. Kang (2005) points out the misogynistic impulse in Pygmalion's creation of a perfect woman and compares modern narratives of female sex robots, which, unlike the ancient myth, have unhappy endings.
8. Marshall (2017) compares the female replicants of the *Blade Runner* films to Pygmalion's creation.
9. Some interpret Apollodorus *Library* 3.14.3 to suggest that a son, Paphos, and a daughter, Metharme, were born to Pygmalion's living statue. Similarly, the plot of *Blade Runner 2049* turns on the magical existence of two children, a girl and a boy who is an exact copy, born to the replicant Rachael, who died in childbirth. See chapter 8 for a Roman-era fantasy about the offspring of the ancient replicant female Pandora.
10. Pygmalion: Ovid *Metamorphoses* 10.243–97; Heraclides Ponticus (lost work) cited by Hyginus *Astronomica* 2.42; Hyginus *Fabulae* 142; Philostephanus of Styrene cited in Clement of Alexandria *Protepticus* 4; *Arnobius Against the Heathen* 6.22. Hansen 2004, 276. Hersey 2009, 94. Reception of Pygmalion myth, Grafton, Most, and Settis 2010, 793–94; Wosk 2015.
11. Raphael 2015, 184–86.
12. Hersey 2009. "Pygmalionism" differs from statue lust; it requires a lover to mimic a statue and then come to life.
13. Philostratus *Lives of the Sophists* 2.18.
14. Homer *Iliad* 2.698–702 and commentary at 701 by Eustathius; Apollodorus *Epitome* 3.30; Ovid *Heroides* 13.151; Hyginus *Fabulae* 104; for other ancient sources, see George Frazer's commentary in the Loeb ed. of Apollodorus *Epitome*, pp. 200–201n1.
15. Wood 2002, 138–39. Hersey 2009, 90–97. Athenaeus *Learned Banquet* 13.601–606; citing the poets Alexis, Adaeus of Mytilene, Philemon, and Polemon. Truitt 2015a, 101.

16. Scobie and Taylor 1975, 50. Hersey 2009, 132. Cohen 1966, 66–67. Innovations in art evoked awe in antiquity, D'Angour 2011, 148–56. An early prototype is Harmony, a realistic AI sexbot from Abyss Creations, made for sex and "companionship," Maldonado 2017. On sex robots, see Devlin 2018.
17. The Tocharian version (sixth to eighth century AD) of a lost Sanskrit text of unknown date, translated by Lane (1947, 41–45). For Hindu and Buddhist automata, see Cohen 2002, 70–71, for discussion of this tale. See also Raghavan 1952; Ali 2016.
18. Cohen 2002, 69, 71, original italics. On Buddhism and robots, Simons 1992, 29–31; Buddhism and biotechnology, see essay by David Loy in Walker 2000, 48–59; on Buddhism and robots, see Mori 1981 and 2012; Borody 2013. On Chinese Buddhism and replicas, Han 2017. On Buddhist perspectives on robots and AI, see Lin, Abney, and Bekey 2014, 69–83.
19. Kang 2011, 15–16; Kang does not address the ancient literary and artistic evidence for Prometheus's construction of the first humans using artisans' tools and methods.
20. The differences between Neoplatonism and Christianity were expounded by the Church Father Tertullian, who was active in the third century AD when these sarcophagi were made. Raggio 1958, 46–50 and figs. Tertullian *Apologeticum* 18.3. Roman mosaic of Prometheus creating the first man, Shahba, Syria, third century AD. Roman sarcophagus showing Prometheus with first man lying at his feet, fourth century AD, Naples museum. See Tassinari 1992 on Neoplatonic, Pythagorean, Orphic, Christian, and Gnostic links to Prometheus as creator.
21. Simons 1992, 24–28, also contrasts Pygmalion and Prometheus.
22. I am grateful to Gabriella Tassinari for discussing the difficulties of determining the dates (and authenticity) of the gems in her catalogue and in other museum collections. For each gem discussed and illustrated in this chapter, see the sources for dating cited in Tassinari 1992; 75–76 for Prometheus working on the form of a woman. I thank Erin Brady for providing an English translation of Tassinari's monograph.
23. Raggio 1958, 46. Apollodorus *Library* 1.7.1; Pausanias 10.4.4. Tassinari 1992, 61–62, citing works by Philemon, Menander, Erinna, Callimachus, Apollodorus, Aesop, Ovid, Juvenal, and Horace referring to Prometheus as the creator of man. See chapter 4, on Prometheus's concerns for the vulnerable human race.
24. Ambrosini 2014; Richter 2006, 53, 55, 97; Dougherty 2006, 17. De Puma 2013, 283. *LIMC* 7 (Jean-Robert Gisler). Spier 1992, 70, 87, nos. 144 and 200, for examples and bibliography. Craftsmen and artisans on Etruscan gems, Ambrosini 2014; for artisans working on herms or busts, 182. Larissa Bonfante, per. corr. March 11, 2017. The customers who owned the gems like those in figs. 6.3–6.11 may have been fellow craftsmen taking pride in their craft, Tassarini 1992.
25. Tassinari 1992, 73–75, 78–80. The antiquity of the gems in figs. 6.3 and 6.4 is not in doubt.
26. Gems showing Prometheus assembling the first man are catalogued by Tassinari (1992). Hatched borders, as in figs 6.7 and 6.10, were favored by Etruscan engravers. Richter 2006, 48, 53, 55, on 97 notes that gem no. 437, plate 14, is not a warrior with a mutilated body because the decapitated head and limbs are not included;

compare Boston Museum of Fine Arts, third century BC, Etruscan gem acc. no. 23.599, depicting *maschalismos*, with two warriors with weapons hacking up an enemy's body. *Maschalismos*, Tassinari 1992, 72; and De Puma 2013, 280–95, esp. 286, discussion of gem no. 7.100. Ambrosini 2014, 182–85, Etruscan gems depicting sculptors working on herms, busts, and statues of women.

27. The exceptional imagery of the second type of gems leads some scholars to question whether some could be neoclassical copies. Thanks to Laura Ambrosini, Ulf Hansson, Ingrid Krauskopf, Claire Lyons, Gabriella Tassinari, and Jean Turfa for discussion and bibliography. Martini 1971, 111, cat. no. 167, pl. 32,5; Krauskopf 1995; Ambrosini 2011, 79, no. 5, fig. 126a–c and bib. Tassinari 1992, 81–82.
28. Carafa 1778, 5–6, plate 23, for the engraving of the first gem with horse and ram; see Scarisbrick, Wagner, and Boardman 2016, 141, fig. 129, for the quoted text, color photos of the gem, ring, and cast, now in the Beverley Gem Collection, Alnwick Castle, United Kingdom. See also Tassinari 1992, 78–79. Skeletons rare in art, Dunbabin 1986.
29. The dates of figs. 6.7 and 6.10 are unresolved (numbers 63 and 54, respectively, in Tassinari 1992 catalogue; figs. 6.8 and 6.10 were not analyzed by Tassinari in 1992; fig. 6.11 (number 59 in Tassinari 1992) is certainly ancient. Thanks to Gabriella Tassinari, personal communications, January–February 2018.
30. Richey 2011, quote 194, 195–96, 202–3; Needham 1991, 2:53–54; Liu 2011, 243–44. Cf. Ambrosino 2017 on the innards of cyborg humanoids.
31. Mattusch 1975, 313–15.
32. Mattusch 1975, 313–15; Aristotle *History of Animals* 515a34–b; cf. *Generation of Animals* 743a2 and 764b29–31; *Parts of Animals* 654b29–34. See De Groot 2008 on Aristotle and mechanics. Cf. Berryman 2009, 72–74, who argues that Aristotle's language is not mechanistic.
33. Cohen 2002, 69. On free will, see Harari 2017, 283–85.
34. The pioneer of Artificial Intelligence, Alan Turing, devised a test in 1951 to reveal whether a machine is sentient, Zarkadakis 2015, 48–49, 312–13. See also Cohen 1963 and 1966, 131–42; Mackey 1984; Berryman 2009, 30; Kang 2011, 168–69. Since Turing, other AI-human tests have been developed: Boissoneault 2017. Paranoid sci-fi themes of androids and false selfhood, Zarkadakis 2015, xv, 53–54, 70–71, 86–87.
35. Boissoneault 2017; Zarkadakis 2015, 36–38, 112–15.
36. Mackey 1984; Gray 2015; Mendelsohn 2015; Shelley 1831 [1818]; Weiner 2015; Cohen 1966; Harari 2017.
37. Dougherty 2006. Note that this Athenian torch race honoring Prometheus had nothing to do with the ancient Olympic Games. The modern Olympic torch relay was introduced by the Nazis for the Berlin Olympics, 1936.
38. Raggio 1958, e.g., 50–53. Reception of Prometheus, see Grafton, Most, and Settis 2010, 785.
39. Godwin's *Lives of the Necromancers* was published in 1834. Galvanism experiments and Shelley's other influences: Zarkadakis 2015, 38–40; Hersey 2009, 106, 146–50;

Kang 2011, 218–22. Zarkadakis 2015, 63–66. Frightening robots figure in E.T.A. Hoffman's German short stories from Shelley's time, "The Automata" (1814) and "The Sandman" (1816) about a wax automaton named Olympia: Cohen 1966, 61–62.

40. Florescu 1975. A striking feature of the 1931 Karloff monster, the two metal bolts on his neck representing crude electrodes, placed on his jugular veins, bringing to mind the placement of the metal bolt on the ankle of the bronze robot Talos (chapter 1). See chapter 9 for the primitive electrical "Baghdad batteries." Kant, "The Modern Prometheus," Rogers and Stevens 2015, 3, and on Shelley's Frankenstein, 1–4. Weiner 2015, 46–74.
41. Prometheus making the first humans was a favored theme in "antiquarian" neoclassical gems carved by European craftsmen in the seventeenth to nineteenth century, collected by Tassie and Prince Poniatowski; Tassinari 1996.
42. Shelley and Lucan: Weiner 2015, 48–51, 64–70; Lucan *Civil War* 6.540–915. On Egyptian demotic tales of necromancy, Mansfield 2015. On mechanical motion eliciting the Uncanny Valley reaction, Zarkadakis 2015, 69; Mori 2012.
43. Shelley 1831. Raggio 1958. Quote, Simons 1992, 27–28. Rogers and Stevens 2015, 1–5.
44. Hyginus *Astronomica* 2.15, *Fabulae* 31, 54, 144.
45. David-Neel 1959, 84.
46. Tales of artificial flying birds appear in ancient Hindu and Mongolian literature too, including a pair of mechanical swans (*yantrahamsa*) "programmed" to steal royal jewels and a legendary Garuda bird that was "steered by pins and pegs." Cohen 2002, 67–69.

CHAPTER 7. HEPHAESTUS: DIVINE DEVICES AND AUTOMATA

1. For the smith god in ancient literature and art, Gantz 1993, 1:74–80. Hephaestus's father was Zeus according to Homer, but he had no father according to Hesiod. For the works of Hephaestus, Pollitt 1990, 15–18. Prosthetic limbs and replacement body parts as artificial human enhancements, chapter 4. Zarkadakis 2015, 79–80.
2. Paipetis 2010 and Vallianatos 2017. On the vivid, kinetic descriptions of Achilles's shield in Homer, in which an "impossible" object is described with hyperrealism and movement, see Francis 2009, 6–13. See also Kalligeropoulos and Vasileiadou 2008.
3. Homer *Iliad* 18.136, 18.368–72, 19.23. "Artificial world," Raphael 2015, 182.
4. Francis 2009, 11–13.
5. Bronze cuirasses and greaves with delineated musculature were used from the sixth century BC on, with many examples recovered from archaeological excavations. Steiner 2001, 29. Other warrior cultures, such as Rome, India, and Japan, also wore anatomical cuirasses.
6. On a fresco from Pompeii, first century AD, Hephaestus, surrounded by tools and half-finished projects, shows Thetis the shield he has made for Achilles.

7. Homer *Iliad* 5.745–50; Mendelsohn 2015, 1.
8. The net, Homer *Odyssey* 8.267ff. Hera's special chair in literature and art, Gantz 1993, 1:75–76.
9. Argus Panoptes: Hesiod *Aegimius* frag. 5. Apollodorus *Library* 2.1.2; Ovid *Metamorphoses* 1.264. Many-eyed Argus appears on a red-figure hydria, fifth century BC, Museum of Fine Arts, Boston, Lefkowitz 2003, 216–17 fig. Argus Painter name vase, stamnos, 500–450 BC, Vienna Kunsthistorisches Museum 3729; Meleager Painter krater, 400 BC, Ruvo Museo Jatta 36930; another double-headed Argus, black-figure amphora, 575–525 BC, British Museum B164. The Pan Painter vase with janiform head and eyes: Misailidou-Despotidou 2012.
10. Soldiers and sleep: Lin 2012, 2015; Lin et al. 2014.
11. On modern "black box" technology inscrutable to users and makers, see introduction and Knight 2017.
12. Apollodorus *Epitome* 5.15–18. *LIMC* 3,1:813–17. According to Bonfante and Bonfante 2002, 202, Pecse is the Etruscan name for the Trojan Horse.
13. Bonfante and Bonfante (2002, 198) suggest that Etule is the Etruscan name for Aetolus, who was confused with his brother Epeius, maker of the Trojan Horse. Metapontum founded by Epeius and his tools displayed in the Temple of Athena: Pseudo-Aristotle *On Marvelous Things Heard* 840A.108, "in the district called Gargaria, near Metapontum, they say that there is a temple of the Hellenian Athene where the tools of Epeius are dedicated, with which he made the wooden horse.... Athena appeared to him in a dream and demanded that he should dedicate the tools to her." Per Justin 20.2, Metapontum was founded by Epeius, the hero who constructed the wooden horse at Troy; in proof of which the inhabitants showed his tools in the Temple of Athena/Minerva.
14. De Grummond 2006, 137–38, fig. VI.31. Images of blacksmiths, craftsmen, and Sethlans on Etruscan gems, Ambrosini 2014, 177–81. Plaster or clay molds for bronze casting, Konstam and Hoffmann 2004. Athena making clay horse, Cohen 2006, 110–11. Another vase painting shows Athena constructing the Trojan Horse, kylix by the Sabouroff Painter, fifth century BC, Archaeological Museum, Florence.
15. Apollodorus *Library* 2.4.7–7, 3.192; Hyginus *Fabulae* 189 and *Astronomica* 2.35; Ovid *Metamorphoses* 7.690–862; Pausanias 9.19.1.
16. Pausanias 10.30.2; Antoninus Liberalis *Metamorphoses* 36 and 41. Telchines and Dactyles associated with animated statues, Blakely 2006, 16, 24, 138, 159, 203, 209, 215–23. Kris and Kurz 1979, 89. Golden Hound versions: Faraone 1992, 18–35; Steiner 2001, 117. See chapter 8 for Pandora, who was made of clay, yet later authors could not resist claiming that she gave birth to offspring. A similar "miracle" is the theme in the 2017 film *Blade Runner 2049*.
17. Faraone 1992, 18–19, 29n1. Marconi 2009.
18. Faraone 1992, 19–23, 13n8. *Pharmaka* "animates" the statues with a kind of "soul" or life but does not necessarily make them move. Hollow statues as vessels that are vivified by being filled with substances, Steiner 2001, 114–20.

19. Asimov's laws, Kang 2011, 302. Future of Life Institute's Beneficial AI Conference 2017; FLI's board included Stephen Hawking, Frank Wilczek, Elon Musk, and Nick Bostrom. https://futurism.com/worlds-top-experts-have-created-a-law-of-robotics/. See also Leverhulme Centre for the Future of Intelligence: http://lcfi.ac.uk/.
20. Martinho-Truswell 2018.
21. Four-wheeled carts, Morris 1992, 10. A small, shallow bronze basin-cart on three wheels, an ancient example of *pen*, bonsai basin, was excavated in a sixth/fifth century BC archaeological site in China, indicating that the idea of a wheeled tripod was put into practice elsewhere in antiquity, Bagley et al. 1980, 265, 272, color plate 65. Photo and explanation of the replica of Hephaestus's wheeled tripod, Kotsanas 2014, 70. the museum is in Katakolo, near Pyrgos, Greece: http://kotsanas.com/gb/index.php.
22. See chapter 9 for more automata in the form of humans and animals made by Philo; for diagrams and photos of a working model of the wine servant, Kotsanas 2014, 52–55.
23. Truitt 2015a, 121–22, plate 27. Badi' az-Zaman Abu I-Izz ibn ar-Razaz al-Jazari (AD 1136–1206): Zielinski and Weibel 2015, 9.
24. Homer *Iliad* 18.360–473. Pasiphae's cow and the Trojan Horse were also mounted on wheels in literature and art. On Hephaestus, his forge and automata, Paipetis 2010, 95–112.
25. Diodorus Siculus 9.3.1–3 and 9.13.2; Plutarch *Solon* 4.1–3.
26. Berlin Painter, Attic hydria from Vulci, ca. 500–480 BC; the quote comes from the Vatican Museum text, cat. 16568; Beazley archive 201984. The priestess seated on the tripod of the Delphic oracle appears on an Attic kylix by the Kodros Painter, from Vulci, ca. 440 BC, Berlin inv. F 2538.
27. Hephaestus in the winged chair decorated with crane's head and tail on an Attic red-figure kylix attributed to the Ambrosios Painter, Berlin 201595, now lost. Triptolemus in his winged chariot with two serpent heads and tails appears in several ancient vase paintings, e.g., a skyphos of about 490–480 BC attributed to Makron, British Museum E140, Beazley 2014683. The Berlin Painter's stamnos showing Triptolemus in his flying chair, ca. 500–470 BC, is in the Louvre inv. G371; the Berlin Painter's kylix with Triptolemus is in Museo Gregoriano Etrusco, Vatican Museums. On the winged chairs, see Matheson 1995b, 350–52.
28. Only a fragment of Pindar's poem survives, Faraone 1992, 28 and 35n86. Marconi 2009.
29. Mendelsohn 2015.
30. Steiner 2001, 117. Francis 2009, 8–10; the Golden Maidens are neither real humans nor inert matter, and so belong in a unique category of being, 9n23.
31. Raphael 2015, 182. Human-computer interface and thought-controlled machines, Zarkadakis 2015; "The Next Frontier: When Thoughts Control Machines" 2018. The Golden Maidens would appear to be Type III AI; see glossary. On black box dilemmas, see "AI in Society: The Unexamined Mind" 2018.

32. Mendelsohn 2015. Cf. Paipetis 2010, 110–12.
33. Big data, AI, and machine learning, Tanz 2016; see also Artificial Intelligence, "general AI," in the glossary.
34. "Magic is linked to science in the same way as it is linked to technology. It is not only a practical art, it is also a storehouse of ideas," Blakely 2006, 212. Maldonado 2017 reports that the sex robot-companion called "Harmony," made by Realbotix for Abyss Creations, was endowed with a "data dump": she is programmed with about five million words, the entirety of Wikipedia, and several dictionaries.
35. Valerius Flaccus *Argonautica* 1.300–314. Paipetis 2010. LaGrandeur 2013, 5. Homer *Odyssey* 8.267. In Hindu texts and Sanskrit epics, Vimāna is a flying palace or chariot controlled by the mind. A fleet of intelligent ships controlled by "the mind or minds" figures in *The Culture* science-fiction series (1987–2012) by Iain M. Banks; thanks to Ingvar Maehle for this reference. The Phaeacian ships appear to be Type III AI; see glossary.
36. Mansfield 2015, 8–10; Lichtheim 1980, 125–51; and Raven 1983 on magical, realistic, and animated wax figures in Egyptian texts and archaeological examples.
37. Paipetis 2010, 97–98.
38. On the ancient human impulse to automate tasks and tools to save labor and improve on human abilities, Martinho-Truswell 2018. The automatic bellows appear to be Type II AI; see glossary.
39. Aristotle's comment (1253b29–1254a1) that self-animated devices could perform slave's work, fitting the "economic" function of robots, suggests that the invention of such devices would abolish slavery. John Stuart Mill (1806–1873) studied Aristotle; it is interesting to compare his statement about automaton workers in *On Liberty* to Aristotle's remarks: "Supposing it were possible to get houses built, corn grown, battles fought, causes tried, and even churches erected and prayers said, by machinery—by automatons in human form," writes Mill. It would be a shame to replace with automatons "the men and women who at present inhabit the more civilized parts of the world, and who assuredly are but starved specimens of what nature can and will produce." After all, "human nature is not a machine to be built after a model, and set to do exactly the work prescribed for it." It is the nature of living things to "grow and develop," and humankind should concentrate on "perfecting and beautifying" human beings themselves. Thanks to Ziyaad Bhorat for bringing this passage to my attention. See Walker 2000 for prescient essays on the dangers of newly emerging genetic engineering and biotechnology. See Bryson 2010 for the caution that robots and AI ought to remain "slaves" of humans.
40. Mendelsohn 2015. LaGrandeur 2013, 9–10. *Robota* derives from Slavic words for drudgery and medieval servitude, Kang 2011, 279; on robot rebellion, 264–96. Čapek, see Simons 1992, 33. Rogers and Stevens 2015. Walton 2015.
41. Berryman 2009, 22, 24–27. Berryman's earlier 2003 paper mentioned Talos.
42. Truitt 2015a, 3–4, the duties of Hephaestus's twenty tripods are conflated with those of the golden assistants.
43. Kang 2011, 15–22.

CHAPTER 8. PANDORA: BEAUTIFUL, ARTIFICIAL, EVIL

1. *Dolos*, trick, snare, trap; Hesiod *Theogony* 589; *Works and Days* 83. "Mr. Afterthought," Faraone 1992, 104.
2. Pandora in ancient art and literature, Gantz 1992, 1:154–59, 162–65; Hard 2004, 93–95; Shapiro 1994, 64–70; Panofsky and Panofsky 1991; Reeder 1995, 49–56; Glaser and Rossbach 2011. Hesiod *Works and Days* 45–58 and *Theogony* 560–71, *kalon kakon* 585; Aeschylus frag. 204; Hyginus *Fabulae* 142 and *Astronomica* 2.15; Sophocles's lost play *Pandora*; Babrius *Aesop's Fables* 58. Reception of Hesiod and the Pandora myth, Grafton, Most, and Settis 2010, 435–36, 683–84.
3. Early Christian writings compare Pandora and Eve: Panofsky and Panofsky 1991, 11–13.
4. Morris 1992, 32–33; Steiner 2001, 25–26, 116–17, 186–90; Francis 2009, 13–16; Brown 1953, 18; Mendelsohn 2015; Lefkowitz 2003, 25–26.
5. Morris 1992, 30–33, 230–31. Francis 2009, 14.
6. Steiner 2001, 116, Hesiod in the *Theogony* presents Pandora as "nothing more than a compilation of her clothing and adornment"; while in *Works and Days* she is composed of interior attributes as well. Faraone 1992, 101.
7. Steiner 2001, 191n25. Hesiod's language and similes "draw attention simultaneously to the vividness and vigor" of this "fabricated living statue" and to the fact that she "is a representation, not the 'real' thing. Why use this language" otherwise? Pandora is the "first manufactured identity"; she is "quite literally built . . . not a product of nature." Francis 2009, 14. Cf. Faraone 1992, 101–2.
8. Faraone 1992, 102–3, discusses Pandora's creation as an animated statue. On alternative versions claiming that Prometheus was the maker of the first woman, see Tassinari 1992, 75–76.
9. On myths describing the Trojan Horse as an animated statue and ancient "tests" to determine whether it and other realistic statues were real or artificial, Faraone 1992, 104–6. Turing test and the like: Kang 2011, 298; Zarkadakis 2015, 48–49, 312–13; Boissoneault 2017.
10. Hesiod's poems do not mention offspring. As they did for Pygmalion's Galatea (see chapter 6), later writers embellished the myth by giving Pandora a daughter by Epimetheus, Pyrrha, wife of Deucalion: Apollodorus *Library* 1.7.2; Hyginus *Fabulae* 142; Ovid *Metamorphoses* 1.350; Faraone 1992, 102–3. No myths recount Pandora's death. Pandora is "outside the natural cycles": Steiner 2001, 187.
11. Raphael 2015, quote 187; compare Steiner 2001, 25. Plato *Laws* 644e on human agency and chapter 6.
12. Mendelsohn 2015. Faraone 1992, 101. On the similarities between Pandora and the golden servants of Hephaestus, Francis 2009, 13. Pandora does not speak in any surviving myths.
13. For ancient representations of Pandora in Italy, Boardman 2000.
14. Reeder 1995, 284–86.

15. Gantz 1993, 1:163–64; Shapiro 1994, 69; Neils 2005, 38–39. Satyrs with hammers, Polygnotus Group vase, Matheson 1995a, 260–62. Penthesilea Painter vase, Boston Museum of Fine Arts 01.8032.
16. Neils 2005, 39. The sown army of automaton soldiers also rose from the earth, chapter 4.
17. Gantz 1993, 1:157–58 and n12; Mommsen in CVA Berlin V, pp. 56–59, Tafel 43, 3–4, and Tafel 47, 6, citing Panofka. Thanks to David Saunders for valuable discussion of this vase. For Etruscan gems depicting Prometheus or Hephaestus working on a small female figure in their laps, see Tassinari 1992, 75–76.
18. Reeder 1995, 281 (quote); 279–81.
19. Shapiro 1994, 66.
20. Steiner 2001, 116–17.
21. As far as I know, this intriguing border pattern on the Niobid and Ruvo kraters has not been noticed by scholars. The British Museum calls it a "dart and lotus" design; others have referred to a slightly similar motif as "Lesbian kyma." A variation of this design appears on the volute kraters Naples H2421 and Bologna 16571 attributed to the Boreas Painter, ca. 480 BC. The design on the Niobid Painter's Pandora vase appears to more strongly represent blacksmith's tongs or an artisan's compass (fabled to have been invented by Daedalus or his nephew Talos). Some also point out that it could represent a blacksmith's bellows. I thank Bob Durrett, Steven Hess, Fran Keeling, David Meadows, and David Saunders for discussing this border design with me.
22. Shapiro 1994, 67. The frieze below Pandora on the Niobid Painter's vase depicts dancing satyrs, suggesting an association with Sophocles's lost satyr play about Pandora. See also Reeder 1995, 282–84. Pandora holds a wreath or leafy branch in each hand.
23. The Geta Vase is in Agrigento, Sicily; the Niobid massacre krater is in the Louvre.
24. Rarity and meaning of frontal faces and emotions on vases, Korshak 1987; Csapo 1997, 256–57; Hedreen 2017, 163 and n17.
25. The archaic smile appears on the face of a dying warrior on the Temple of Aphaia, Aegina, Greece, and on the face of Antiope being abducted by Theseus, Temple of Apollo, Eretria.
26. The screenplay was written by Lang and his wife, Thea von Harbrou, based on her novel of 1924. Simons 1992, 185; Dayal 2012; Kang 2011, 288–95; Zarkadakis 2015, 50–51.
27. The female robot in *Metropolis* is capable of becoming a simulacrum of Maria. The actress Brigitte Helm was born in 1906; filming began in 1925.
28. Description of the evil fembot, by the actor who played the "mad scientist," Klein-Rogge 1927.
29. Shapiro 1994, 65.
30. Harrison 1999, 49–50.
31. The Pergamon copy of Phidias's Athena and base is in the Pergamon Museum, Berlin. The small replica, the Lenormant Athena and base, is in the National Archaeological

Museum, Athens. Other small Roman copies also exist. Fragments of the marble Pandora frieze and "strange" smiling woman's head: Neils 2005, 42–43, fig. 4.13.

32. Pandora's *pithos* was metal, not earthenware: Neils 2005, 41. Pandora myth in post-classical art and literature: Panofsky and Panofsky 1991, mistranslation, 14–26. Pandora in the arts: Reid 1993, 2:813–17.
33. In later variants of the story, the forbidden jar comes into Epimetheus's possession by other means or is opened by him instead of Pandora, e.g., in Philodemus, first century BC, and Proclus, fifth century AD, Panofsky and Panofsky 1991, esp. 8 and nn11–12.
34. Neils 2005, 40. This pair of *pithoi* reflects the dual positive and ominous uses of large jars in antiquity, for storing food and other vital commodities and as coffins for burying poor folk. Confusingly, two writers of the sixth century BC, Theognis frag. 1.1135 and Aesop *Fables* 525 and 526/Babrius 58, claimed that Pandora brought Zeus's jar of blessings to earth and that Elpis/Hope was a positive thing in that urn; see discussion below.
35. British Museum 1865,0103.28: Neils 2005, 38–40 and figs. 4.1–2 and 4.6–8. *LIMC* 3, s.v. Elpis, no. 13; Reeder 1995, 51 fig. 1–4.
36. Neils 2005, 41–42.
37. The Early Christian Father Origen (b. AD 185) found the pagan myth of Pandora "laugh-provoking," Panofsky and Panofsky 1991, 12–13; see 7n12 for Macedonius Consul's cynical epigram (sixth century AD) that begins, "I smile when I look at Pandora's jar."
38. Harrison 1986, 116; Neils 2005, 43.
39. Gantz 1993, 1:157. Aesop (*Fables* 525 and 526, early sixth century BC) wrote that a jar of Good Things had been entrusted to mankind by Zeus, "but man had no self-control and he opened the jar—all the Good Things flew out." They were chased away by the stronger evils in the world, and flew back up to Olympus to reside with the gods. Now they are doled out to humans one at a time, to "escape notice of the Evil Things which are ever-present. Hope remained in the jar, however, the one Good Thing left to humankind to console them with the promise of the Good Things we have lost." In the late sixth century BC, Theognis (*Elegies*) tells a similar tale, remarking that hope was the "only deity left on earth, for the rest have flown." Aesop and Theognis agree with Hesiod that Hope alone stayed behind, and they view Hope in a positive light.
40. Fairy-tale versions, Panofsky and Panofsky 1991, 110–11. Aristotle *On Memory* 1.449b25–28.
41. According to Plato (*Gorgias* 523a), it was Zeus who told Prometheus to deprive men of the foreknowledge of death. In *Protagoras* 320c–322a, Plato refers indirectly to Epimetheus's mistake.
42. Thanks to Josiah Ober for help in setting up a standard two-by-two, four-box matrix with rows designated "good" and "evil" and columns "activated" and "unactivated." For various modern opinions, see, e.g., Hansen 2004, 258; Lefkowitz 2003, 233.
43. Ethical challenges of advancing robotics and AI technologies: Lin, Abney, and Bekey 2014, 3–4, the qualms about automata and human enhancement via technology

have very deep roots, going back to antiquity, already posing concerns that would anticipate the "cautionary tales" in modern literature "about insufficient programming, emergent behavior, errors, and other issues that make robots unpredictable and potentially dangerous"; 362, "The mere uttering of the word 'robot' opens up a Pandora's box of images, myths, wishes, illusions, and hopes, which humanity has, over centuries, applied to automata."

44. Compare the evil robot Tik-Tok in Sladek 1983. The premise of the android-hosted amusement park of the *Westworld* TV series is that human guests may indulge their darkest fantasies upon the bodies of the androids, whose programming prevents them from harming humans.

CHAPTER 9. BETWEEN MYTH AND HISTORY: REAL AUTOMATA AND LIFELIKE ARTIFICES IN THE ANCIENT WORLD

1. "Black box" technology, Knight 2017. "Relative modernism," Bosak-Schroeder 2016.
2. Berryman 2009, 69–75. James and Thorpe 1994, 200–225. Marsden 1971. Heron of Alexandria acknowledged that some of his automata mechanisms were related to catapults; Ruffell 2015–16.
3. On links between ruthless tyrants and devices, see Amedick 1998, 498.
4. D'Angour 1999, 25; a jocular article juxtaposing historical evidence for human flight with representations in ancient comedy and fiction.
5. Sappho's supposed suicide at the Leucadian cliff was first suggested in the late fourth century BC by the comic playwright Menander (frag. 258 K).
6. *Book of Sui* (AD 636), Needham and Wang 1965, 587; *Zizhi Tongjian* 167 (AD 1044) in abridgment by Ronan 1994, 285. *History of the Northern Dynasties* 19. James and Thorpe 1994, 104–7 on man-bearing kites and parachutes. Yuan Hangtou survived but was executed.
7. Lucian *Phalaris*. Phalaris's reputation for cruelty: Aristotle *Politics* 5.10; *Rhetoric* 2.20. Pindar *Pythian* 1; Polyaenus *Stratagems* 5.1; Polybius 12.25. Kang 2011, 94–95. Phalaris's sadism was exaggerated by the early Christian writer Tatian, b. AD 120, who claimed that Phalaris devoured infants (*Address to the Greeks* 34).
8. Diodorus Siculus 9.18–19. Plutarch *Moralia* 315. Lucian *Phalaris*.
9. Plutarch *Moralia* 315c–d, 39, citing Callimachus *Aetia* (fourth century BC, known only from fragments, and Aristeides of Miletus's *Italian History* book 4 (lost). See also Stobaeus *Florilegium*, fifth century AD. Arruntius's bronze horse recalls some descriptions of the Trojan Horse, hollow with an opening in the side.
10. Diodorus Siculus 9.18–19 and 13.90.3–5; Cicero *Against Verres* 4.33 and *Tusculan Disputations* 2.7; 5.26, 5.31–33 (death of Phalaris), 2.28
11. *Consularia Caesaraugustana*, the chronicle of Zaragoza, *Victoris Tunnunnensis Chronicon*, ed. Hartmann, Victor 74a, 75a, p. 23, commentary pp. 100–101. For sadistic public displays of roasting birds and animals alive in China, Tang dynasty, for the pleasure of Empress Wu Zetian, see Benn 2004, 130.

12. Berryman (2009, 29–30) includes the Brazen Bull in the "homunculus"-driven variety of artifices in her classification system. For Indian automata worked by people inside, Cohen 2002, 69.
13. Faraone 1992, 21. Blakely 2006, 16, 215–23. The Antikythera device is in the National Archaeological Museum, Athens. Iverson 2017.
14. Faraone 1992, 21, 26. Timaeus in scholia to Pindar *Olympian* 7.160.
15. A drawing of the stentorophonic tube is preserved in the Vatican Museums; see Kotsanas 2014, 83. Stoneman 2008, 121, Aristotle tells Alexander about the "pneumatic horn of Yayastayus," the Horn of Themistius, a "war organ" believed to have been invented ca. AD 800–1100, perhaps powered by pneumatics or hydraulics.
16. Musical automata: Zielinski and Weibel 2015, 49–99. Pollitt 1990, 89.
17. Cohen 1966, 21–22 and n20; other speaking statues, 18–24. Chapuis and Droz 1958, 23–24.
18. Cohen 1966, 15–16. Philostratus *Life of Apollonius of Tyana* 6.4; *Imagines* 1.7. "The Sounding Statue of Memnon" 1850.
19. Cohen 1966, 24; McKeown 2013, 199; LaGrandeur 2013, 22. Himerius *Orations* 8.5 and 62.1.
20. Oleson 2009, 785–97 for Greek and Roman automata. Poulsen 1945; Felton 2001, 82–83.
21. Frood 2003; Keyser 1993, for experiments, diagrams, and photos. The theory that the batteries were used to electroplate silver has been discarded. Thanks to Sam Crow for pointing out that if thin wires once existed, they may have corroded away.
22. Brunschwig and Lloyd 2000, Archytas: 393, 401, 403, 406, 926–27, 932–33; ancient mechanics: 487–94. Keyser and Irby-Massie 2008, 161–62; D'Angour 2003, 108, 127–28, 180–82.
23. Chirping bird devices: Kotsanas 2014, 51 and 69. Sources for Archytas: Aristotle *Politics* 8.6.1340b25–30; Horace *Odes* 1.28; D'Angour 2003, 180–82, ; Plutarch *Marcellus* 14.5–6. Diogenes Laertius 8.83; Aulus Gellius *Attic Nights* 10.12.9–10; Vitruvius *On Architecture* 1.1.17; 7.14. Berryman 2009, 58 and n14, 95 n159 (Aristotle and Archytas); 87–96, Berryman speculates that the "dove" was a nickname for a catapult or projectile, but neither would account for the "current of air and weights" said to propel the flying device. Aulus Gellius's source, Favorinus, a philosopher and historian who was also a friend of Plutarch, wrote nearly thirty works, most known from fragments.
24. See Brunschwig and Lloyd 2000, 933; D'Angour 2003, 181. Huffman 2003, 82–83, 570–78 (dove); for a working aerodynamic replica of Archytas's Dove using a pig's bladder and compressed air or steam, see Kotsanas 2014, 145. The Dove is placed in the category of "mythic self-moving devices of human creation" by Kang 2011, 16–18.
25. Aristotle *Politics* 5.6.1340b26; Huffman 2003, 303–7 (clapper).
26. Plutarch *Demetrius*.; Diogenes Laertius 1925b78.

27. Demochares's history of his times is lost but quoted by Polybius 12.13. D'Angour 2011, 164. Berryman 2009, 29–30.
28. Koetsier and Kerle 2015, fig. 2a and b. The Giant Snail and problems with Rehm's theory, see Ian Ruffell's University of Glasgow blog post "Riding the Snail," March 31, 2016, http://classics.academicblogs.co.uk/riding-the-snail/.
29. Snails in Greek folklore, Hesiod *Works and Days* 571; Plautus *Poen.* 531; Plutarch *Moralia* 525e. Donkeys (asses): Homer *Iliad* 11.558; Simonides 7.43–49; Plautus *Asinaria*; Apuleius *Golden Ass*; etc.
30. Diodorus Siculus frag. 27.1.
31. Polybius 13.6–8; Apega 18.17; also 4.81, 16.13, 21.11. Sage 1935. Pomeroy (2002, 89–90 and n51) accepts authenticity of account, 152.
32. Aristotle *Constitution of Athens*, describes the *kleroterion*; for a surviving example, Dow 1937. Demetrius and Mithradates's attempt to surpass him in 88 BC, Mayor 2010, 179–83. Ancient military technology: Aeneas Tacticus; Philo of Byzantium; Berryman 2009, 70–71; Cuomo 2007; Hodges 1970, 145–53, 183–84; Marsden 1971. Archimedes, Plutarch *Marcellus* 14–18; Brunschwig and Lloyd 2000, 544–53; Keyser and Irby-Massie 2008, 125–28.
33. Mayor 2010, 182, 291–92, 193–94. Kotsanas 2014, deus ex machina model, 101.
34. Koetsier and Kerle 2015.
35. Keyser 2016 on the date of the Grand Procession, marriage to Arsinoe II, and the reliability of Callixenus's account, based on *Accounts of the Penteterides*.
36. Koetsier and Kerle 2015. Athenaeus *Learned Banquet* 11.497d; Keyser and Irby-Massie 2008, 496.
37. Philo, Ctesibius, Heron: Hodges 1970, 180–84. Neither Ctesibius nor Philo of Byzantium receives notice in Minsoo Kang's "historical study of the automaton" as a working object and concept in the European imagination. The unparalleled Nysa automaton is relegated to a footnote, and Demetrius's Great Snail and the deadly Apega "robot" of Sparta are also omitted from Kang's categories of actual mechanical automata of human design in antiquity: Kang 2011, 16–18, 332n66 (Nysa); 1. Sylvia Berryman (2009, 116) briefly mentions the possibility that Ctesibius made the Nysa automaton.
38. Zielinski and Weibel 2015, 20–47; Truitt 2015a, 4, 19; Keyser and Irby-Massie 2008, 684–56.
39. Huffman 2003, 575; Philo *Pneumatics* 40, 42. Diagram of the bird-and-snake assemblage, James and Thorpe 1994, 117. For working models of bronze and wood and explanations of the serving woman, the bird and owl, and the Pan and dragon, see Kotsanas 2014, 51–55.
40. Heron: Woodcroft 1851; Keyser and Irby-Massie 2008, 384–87. Ruffell 2015–16.
41. Working models and explanation of the Heracles-and-dragon mechanism, and the automatic theater, James and Thorpe 1994, 136–38; Kotsanas 2014, 58 and 71–75. Anderson 2012 (the first programmable device is often said to be the Jacquard loom of 1800). Berryman 2009, 30 citing Heron *Automata* 4.4.4. Huffman 2003, 575.

Kang (2011) includes Heron's works in his third category of actually constructed automata, 16.

42. Ruffell 2015–16; for more 3-D re-creations and explanations of Heron's self-moving artifices, see the Heron of Alexandria/Automaton Project directed by Ian Ruffell and Francesco Grillo at the University of Glasgow. http://classics.academicblogs.co.uk/heros-automata-first-moves/.
43. Medieval Islamic and European automata: Brunschwig and Lloyd 2000, 410, 490–91, 493–94. Zielinski and Weibel 2015, 20–21; James and Thorpe 1994, 138–40; Truitt 2015a, 18–20. By the tenth century, Arabic translations of the automata designs of Greek inventors such as Philo and Heron were adapted in India; Ali 2016, 468. Strong 2004, 132n17.
44. Needham 1986; 4:156–63 and throughout, on the history of Chinese mechanical engineering and automatic devices. As Forte (1988, 11) points out, not all mechanical innovations in China were transmitted from Europe; some arose from what Needham termed "diffusion stimulus." South-pointing chariot, James and Thorpe 1994, 140–42.
45. Tang inventions, Benn 2004, 52, 95–96, 108–9, 112, 143–44, 167, 271. Empress Wu Zetian's ambition to outdo Asoka: Strong 2004, 125 and n6 sources. Empress Wu was also called Wu Zhao.
46. Keay 2011, 69 and n19, citing R. K. Mookerji, in *History and Culture of the Indian People*, 2:28. Mookerji describes the armored war chariot with whirling clubs or blades as like a "tank"; Keay calls it a "robot" swinging a club; others compare the "machine" to a scythed chariot with spinning blades attached to the wheels.
47. Strong 2004, 124–38. Keay 2011, 78–100; Ali 2016, 481–84.
48. Strong 2004, 132–38; Pannikar 1984; there are other versions in Cambodian and Thai. Higley 1997, 132–33. Cohen 2002. Zarkadakis 2015, 34. "Drew on a rich store of legends," Ali (2016, 481–84) discusses the legend and the date and sources of the *Lokapannatti.*
49. Strong 2004, 132–33. In some versions, the engineer is beheaded by the robot assassin sent to kill Asoka, Higley 1997, 132–33, and Pannikar 1984.
50. Cohen 2002, 73–74. It is assumed that the *Lokapannatti* story was solely influenced by later Byzantine and early medieval automata. For the history of automata and elaborate mechanical wonders, comparable to the fabulous Byzantine "Throne of Solomon," in early medieval India, see Ali 2016, esp. 484 on the circulation of *techne*, and Brett 1954 on the automated Throne of Solomon.
51. Ali 2016, 484.
52. Greco-Buddhist syncretism, McEvilley 2001; Boardman 2015.
53. Asoka and Hellenistic rulers, Hinuber 2010, 263 (Megasthenes). Megasthenes *Indica* fragments; Arrian *Indica* 10. Megasthenes and Deimachus were envoys to Mauryan emperor Chandragupta and his son; Dionysius was Ptolemy's envoy to Asoka. See Arrian *Anabasis* 5; Pliny 6.21; Strabo 2.1.9–14; 15.1.12.
54. Keay 2011, 78–100; McEvilley 2000, esp. 367–70; on Indian technology, 649 and n19. On Asoka's envoys to Hellenistic rulers, Jansari 2011.

55. Legge 1965, 79. Animated Buddhist statues and carts in China, Needham 1986, 159–60, 256–57. On miracle tales of animated Buddhist statues, Wang 2016.
56. Rotating attendants, Needham 1986, 159. Wrathful Vajrapani, Wang 2016, 32 and 27. Daoxuan, Strong 2004, 187–89. Dudbridge 2005. Daoxuan's sacred technology and descriptions of the utopian Jetavana monastery automata in India, Forte 1988, 38–50nn86 and 92; 49–50, one cannot know whether Daoxuan was describing real automata of India that he had heard or read about, but Empress Wu apparently wished to construct physical replicas of those wonders in her shrines.
57. Hsing and Crowell 2005, esp. 118–23. Greek-Indian influences, Boardman 2015, 130–99; Heracles in Buddhist art, 189, 199, figs 116, 118, 122. Relief panel of Heracles in lion skin with sword: British Museum 1970,0718.1.
58. Simons 1992, 29–32. Mori 1981 and 2012. Borody 2013. Han 2017.
59. Borody 2013. Thanks to Ruel Macraeg for telling me about *Mazinger Z* and *18 Bronzemen* and thanks to Sage Adrienne Smith for telling me about the ancient robots in *Laputa: Castle in the Sky*. "Whistlefax" robot by Glorbes (B. Ross), Fwoosh Forums November 13, 2007, http://thefwoosh.com/forum/viewtopic.php?t=12823&start=4380.
60. Berryman 2009, 28 original italics. D'Angour 2011, 62–63, 108–9, 127, 128–33, 180–81.
61. Zarkadakis 2015, xvii, 305.

EPILOGUE. AWE, DREAD, HOPE: DEEP LEARNING AND ANCIENT STORIES

1. An earlier version of parts of this epilogue appeared in *Aeon*, May 16, 2016. On love/hate responses to AI, Zarkadakis 2015.
2. Microsoft's Tay and Zo, Kantrowitz 2017; human bias in AI, Bhorat 2017. Tay's debut and demise: http://www.telegraph.co.uk/technology/2016/03/24/microsofts-teen-girl-ai-turns-into-a-hitler-loving-sex-robot-wit/ .
3. Raytheon: http://www.raytheon.com/news/feature/artificial_intelligence.html.
4. Hawking quote, Scheherazade: Flood 2016. http://www.news.gatech.edu/2016/02/12/using-stories-teach-human-values-artificial-agents. http://realkm.com/2016/01/25/teaching-ai-to-appreciate-stories/. Summerville et al. 2017, 9–10. Scheherazade: R. Burton, trans. and intro by A. S. Byatt. *Arabian Nights, One Thousand and One Nights*.
5. Zarkadakis 2015, 27, 305. Leverhulme Centre for the Future of Intelligence "AI Narrative" project, http://lcfi.ac.uk/projects/ai-narratives/.
6. "Dawn of RoboHumanity": Popcorn 2016,112–13.

BIBLIOGRAPHY

ANCIENT GREEK AND LATIN TEXTS ARE AVAILABLE IN TRANSLATION IN THE LOEB CLASSICAL LIBRARY OR ONLINE, UNLESS OTHERWISE NOTED.

Aerts, Willem J. 2014. *The Byzantine Alexander Poem.* Berlin: De Gruyter.

"AI in Society: The Unexamined Mind." 2018. *Economist,* February 17, 70–72.

Ali, Daud. 2016. "Bhoja's Mechanical Garden: Translating Wonder across the Indian Ocean, circa. 800–1100 CE." *History of Religions* 55, 4:460–93.

Ambrosini, Laura. 2011. *Le gemme etrusche con iscrizioni.* Mediterranea supplement 6. Pisa-Rome: Fabrizio Serra Editore.

——. 2014. "Images of Artisans on Etruscan and Italic Gems." *Etruscan Studies* 17, 2 (November): 172–91.

Ambrosino, Brandon. 2017. "When Robots Are Indistinguishable from Humans What Will Be inside Them?" *Popular Mechanics* (February 15). http://www.popularmechanics.com/culture/tv/a25210/inside-synths-amc-humans/.

Amedick, Rita. 1998. "Ein Vergnügen für Augen und Ohren: Wasserspiele und klingende Kunstwerke in der Antike (Teil I)." *Antike Welt* 29:497–507.

Anderson, Deb. 2012. "Was There Artificial Life in the Ancient World? Interview with Dr. Alan Dorin." *Sydney Morning Herald,* August 28. http://www.smh.com.au/national/education/was-there-artificial-life-in-the-ancient-world-20120827-24vxt.html.

Apollonius of Rhodes. 2015. *Argonautica Book IV.* Trans. and comm. Richard Hunter. Cambridge: Cambridge University Press.

Ayrton, Michael. 1967. *The Maze Maker.* New York: Holt, Rinehart and Winston.

Bagley, Robert, et al. 1980. *The Great Bronze Age of China.* New York: Metropolitan Museum.

Benn, Charles. 2004. *China's Golden Age: Everyday Life in the Tang Dynasty.* Oxford: Oxford University Press.

Berryman, Sylvia. 2003. "Ancient Automata and Mechanical Explanation." *Phronesis* 48, 4:344–69.

——. 2007. "The Imitation of Life in Ancient Greek Philosophy." In *Genesis Redux: Essays in the History and Philosophy of Artificial Life,* ed. Jessica Riskin, 35–45. Chicago: University of Chicago Press.

——. 2009. *The Mechanical Hypothesis in Ancient Greek Natural Philosophy*. Cambridge: Cambridge University Press.

Bhorat, Ziyaad. 2017. "Do We Still Need Human Judges in the Age of Artificial Intelligence?" Transformation, August 9. https://www.opendemocracy.net/transformation/ziyaad-bhorat/do-we-still-need-human-judges-in-age-of-artificial-intelligence.

Birrell, Anne, trans. 1999. *The Classic of Mountains and Seas*. London: Penguin.

Blakely, Sandra. 2006. *Myth, Ritual, and Metallurgy in Ancient Greece and Recent Africa*. Cambridge: Cambridge University Press.

Blakemore, Kenneth. 1980. "Age Old Technique of the Goldsmith." *Canadian Rockhound*, February.

Boardman, John. 2000. "Pandora in Italy." In *Agathos Daimon, Mythes et Cultes: Etudes d'iconographie en l'honneur de Lilly Kahil*, ed. P. Linant de Bellefonds et al., 49–50. Athens.

——. 2015. *The Greeks in Asia*. London: Thames and Hudson.

Boissoneault, Lorraine. 2017. "Are Blade Runner's Replicants 'Human'? Descartes and Locke Have Some Thoughts." *Smithsonian*, Arts and Culture, October 3. https://www.smithsonianmag.com/arts-culture/are-blade-runners-replicants-human-descartes-and-locke-have-some-thoughts-180965097/.

Bonfante, Giuliano, and Larissa Bonfante. 2002. *The Etruscan Language: An Introduction*. Manchester: University of Manchester Press.

Borody, Wayne A. 2013. "The Japanese Roboticist Masahiro Mori's Buddhist Inspired Concept of 'The Uncanny Valley.'" *Journal of Evolution and Technology* 23, 1:31–44.

Bosak-Schroeder, Clara. 2016. "The Religious Life of Greek Automata." *Archiv für Religionsgeschichte* 17:123–36.

Bremmer, Jan. 2013. "The Agency of Greek and Roman Statues: From Homer to Constantine." *Opuscula* 6:7–21.

Brett, G. 1954. "The Automata in the Byzantine 'Throne of Solomon.'" *Speculum* 29:477–87.

Brinkmann, Vinzenz, and Raimund Wünsche, eds. 2007. *Gods in Color: Painted Sculpture of Classical Antiquity*. Traveling exhibition catalogue. Munich: Biering & Brinkmann.

Brown, Norman O., trans. 1953. *Theogony, Hesiod*. Indianapolis: Bobbs-Merrill.

Brunschwig, Jacques, and Geoffrey Lloyd, eds. 2000. *Greek Thought: A Guide to Classical Knowledge*. Cambridge, MA: Harvard University Press.

Bryson, Joanna. 2010. "Robots Should Be Slaves." In *Close Engagements with Artificial Companions*, ed. Yorick Wilks, 63–74. Amsterdam: John Benjamins.

Bryson, Joanna, and Philip Kime. 2011. "Just an Artifact: Why Machines Are Perceived as Moral Agents." In *Proceedings of the Twenty-Second International Joint Conference on Artificial Intelligence*, vol. 2, ed. T. Walsh, 1641–46. Menlo Park, CA: AAAI Press.

Buxton, Richard. 2013. *Myths and Tragedies in Their Ancient Greek Contexts*. Oxford: Oxford University Press.

Carafa, Giovanni, duca di Noja. 1778. *Alcuni Monumenti del Museo Carrafa in Napoli.* Naples. Digitized by Getty Research Institute in 2016: https://archive.org/details/alcunimonumentidoocara.

Carpino, A. A. 2003. *Discs of Splendor: The Relief Mirrors of the Etruscans.* Madison: University of Wisconsin Press.

Cave, Stephen. 2012. *Immortality: The Quest to Live Forever and How It Drives Civilization.* New York: Crown.

Ceccarelli, Marco, ed. 2004. *International Symposium on History of Machines and Mechanisms.* Dordrecht: Kluwer Academic.

Chapuis, Alfred, and Edmond Droz. 1958. *Automata.* Trans. A. Reid. Neuchatel: Griffon.

Cheating Death." 2016. *Economist*, August 13, 7.

Clarke, Arthur C. 1973. "'Hazards of Prophecy: The Failure of Imagination." Rev. ed. of 1962. *Profiles of the Future: An Enquiry into the Limits of the Possible.* London: Gollanz.

Clauss, James, and Sarah Johnston, eds. 1997. *Medea: Essays on Medea in Myth, Literature, Philosophy, and Art.* Princeton, NJ: Princeton University Press.

Cline, Eric, ed. 2010. *Oxford Handbook of the Bronze Age Aegean.* New York: Oxford University Press.

Cohen, Beth. 2006. *The Colors of Clay.* Los Angeles: Getty Museum.

Cohen, John. 1963. "Automata in Myth and Science." *History Today* 13, 5 (May).

———. 1966. *Human Robots in Myths and Science.* London: Allen and Unwin.

Cohen, Signe. 2002. "Romancing the Robot and Other Tales of Mechanical Beings in Indian Literature." *Acta Orientalia* (Denmark) 64:65–75.

Colarusso, John, trans. 2016. *Nart Sagas: Ancient Myths and Legends of the Circassians and Abkhazians.* Princeton, NJ: Princeton University Press.

Cook, A. B. 1914. *Zeus: A Study in Ancient Religion.* Vol. 1. Cambridge: Cambridge University Press.

Cooper, Jean. 1990. *Chinese Alchemy: The Taoist Quest for Immortality.* New York: Sterling.

Csapo, Eric. 1997. "Riding the Phallus for Dionysus: Iconology, Ritual, and Gender-Role De/Construction." *Phoenix* 51, 3/4:253–95.

Cuomo, Serafina. 2007. *Technology and Culture in Greek and Roman Antiquity.* Cambridge: Cambridge University Press.

Cusack, Carole. 2008. "The End of Human? The Cyborg Past and Present." *Sydney Studies in Religion*, September 19, 223–34.

D'Angour, Armand. 1999. "Men in Wings." *Omnibus Magazine* 42 (Classical Association, 2001): 24–25.

———. 2003. "Drowning by Numbers: Pythagoreanism and Poetry in Horace Odes 1.28." *Greece and Rome* 50, 2:206–19.

———. 2011. *The Greeks and the New: Novelty in Ancient Greek Imagination and Experience.* Cambridge: Cambridge University Press.

Darling, Kate, Palash Nandy, and Cynthia Breazeal. 2015. "Empathetic Concern and the Effect of Stories in Human-Robot Interaction." Proceedings of the IEEE

International Workshop on Robot and Human Communication (ROMAN), February 1. https://ssrn.com/abstract=2639689.

David-Neel, Alexandra. 1959. *The Superhuman Life of Gesar of Ling*. London: Rider. Reprint, 2001, Shambhala.

Dayal, Geeta. 2012. "Recovered 1927 *Metropolis* Film Program Goes behind the Scenes of a Sci-Fi Masterpiece." *Wired*, July 12. https://www.wired.com/2012/07/rare-metropolis-film-program-from-1927-unearthed/?pid=7549&pageid=112666&viewall=true.

de Grey, Aubrey. 2008. "Combating the Tithonus Error." *Rejuvenation Research* 11, 4:713–15.

de Grey, Aubrey, with Michael Rae. 2007. *Ending Aging: The Rejuvenation Breakthroughs That Could Reverse Human Aging in Our Lifetime*. New York: St. Martin's Press.

De Groot, Jean. 2008. "Dunamis and the Science of Mechanics: Aristotle on Animal Motion." *Journal of the History of Philosophy* 46, 1:43–68.

de Grummond, Nancy Thomson. 2006. *Etruscan Myth, Sacred History, and Legend*. Philadelphia: University of Pennsylvania Museum.

De Puma, Richard. 2013. *Etruscan Art: In the Metropolitan Museum of Art*. New York: Metropolitan Museum.

Devlin, Kate. 2018. *Turned on: Science, Sex and Robots*. London: Bloomsbury.

Dickie, Matthew. 1990. "Talos Bewitched: Magic Atomic Theory and Paradoxography in Apollonius *Argonautica* 4.1638–88." *Papers of the Leeds International Latin Seminar* 6, ed. F. Cairns and M. Heath, 267–96.

———. 1991. "Heliodorus and Plutarch on the Evil Eye." *Classical Philology* 86, 1:17–29.

Donohue, Alice A. 1988. *Xoana and the Origins of Greek Sculpture*. Oxford.

Dougherty, Carol. 2006. *Prometheus*. London: Routledge.

Dow, Sterling. 1937. "Prytaneis. A Study of the Inscriptions Honouring the Athenian Councillors." *Hesperia* Suppl. 1, Athens: American School of Classical Studies.

Dudbridge, Glen. 2005. "Buddhist Images in Action." In *Books, Tales and Vernacular Culture: Selected Papers on China*, 134–50. Leiden: Brill.

Dunbabin, Katherine. 1986. "Sic erimus cuncti . . . The Skeleton in Greco-Roman Art." *Jahrbuch des Deutsches Archäologischen Instituts* 101 (1986): 185–255.

Eliade, Mircea. 1967. *From Primitives to Zen: A Thematic Sourcebook of the History of Religions*. New York: Harper & Row.

Faraone, Christopher. 1992. *Talismans and Trojan Horses: Guardian Statues in Ancient Greek Myth and Ritual*. Oxford: Oxford University Press.

Felton, Debbie. 2001. "The Animated Statues of Lucian's Philopseudes." *Classical Bulletin* 77, 1:75–86.

Flood, Alison. 2016. "Robots Could Learn Human Values by Reading Stories." *Guardian*, February 18.

Florescu, Radu. 1975. *In Search of Frankenstein: Exploring the Myths behind Mary Shelley's Monster*. New York: Little, Brown.

Forte, Antonio. 1988. *Mingtang Utopias in the History of the Astronomical Clock: The Tower, Statue and Armillary Sphere Constructed by Empress Wu*. Rome: Istituto Italiano per il Medio ed Estremo Oriente.

Francis, James A. 2009. "Metal Maidens, Achilles' Shield, and Pandora: The Beginnings of 'Ekphrasis.'" *American Journal of Philology* 130, 1 (Spring): 1–23.

Friend, Tad. 2017. "The God Pill: Silicon Valley's Quest for Eternal Life." *New Yorker*, April 3, 54–67.

Frood, Arran. 2003. "The Riddle of Baghdad's Batteries." BBC News, February 27. http://news.bbc.co.uk/2/hi/science/nature/2804257.stm.

Gantz, Timothy. 1993. *Early Greek Myth: A Guide to Literary and Artistic Sources*. 2 vols. Baltimore: Johns Hopkins University Press.

Garten, William, and Frank Dean. 1982. "The Evolution of the Talos Missile." *Johns Hopkins Applied Physics Laboratory Technical Digest* 3, 2:117–22. http://www.jhuapl.edu/techdigest/views/pdfs/V03_N2_1982/V3_N2_1982_Garten.pdf.

Glaser, Horst Albert, and Sabine Rossbach. 2011. *The Artificial Human: A Tragical History*. New York: Peter Lang.

Godwin, William. 1876 [1834]. *Lives of the Necromancers*. London. Chatto and Windus. https://archive.org/details/livesnecromance04godwgoog.

Grafton, Anthony, Glenn Most, and Salvatore Settis. 2010. *The Classical Tradition*. Cambridge: Harvard University Press.

Gray, John. 2015. *The Soul of the Marionette*. London: Penguin.

Hales, Thomas C. 2001. "The Honeycomb Conjecture." *Discrete and Computational Geometry* 25, 1:1–22. https://www.communitycommons.org/wp-content/uploads/bp-attachments/14268/honey.pdf.

Hallager, Erik. 1985. *The Master Impression: A Clay Sealing from the Greek-Swedish Excavations at Kastelli, Khania*. Studies in Mediterranean Archaeology 69. Goteburg: Paul Forlag Astroms.

Han, Byung-Chul. 2017. *Shanzhai: Deconstruction in Chinese (Untimely Meditations, Book 8)*. Trans. P. Hurd. Boston: MIT Press.

Hansen, William. 2002. *Ariadne's Thread: A Guide to International Tales Found in Classical Literature*. Ithaca, NY: Cornell University Press.

———. 2004. *Handbook of Classical Mythology*. London: ABC-CLIO.

———. 2017. *The Book of Greek and Roman Folktales, Legends, and Myths*. Princeton, NJ: Princeton University Press.

Harari, Yuval Noah. 2017. *Homo Deus: A Brief History of Tomorrow*. New York: Harper.

Hard, Robin. 2004. *The Routledge Handbook of Greek Mythology*. London: Routledge.

Harrison, Evelyn. 1986. "The Classical High-Relief Frieze from the Athenian Agora." *Archaische und klassische griechische Plastik* 2:109–17.

———. 1999. "Pheidias." In *Personal Styles in Greek Sculpture*, ed. O. Palagia and J. Pollitt, 16–65. Cambridge: Cambridge University Press.

Hawes, Greta. 2014. *Rationalizing Myth in Classical Antiquity*. Oxford: Oxford University Press.

Haynes, Natalie. 2018. "When the Parthenon Had Dazzling Colors." BBC News, January 22. http://www.bbc.com/culture/story/20180119-when-the-parthenon-had-dazzling-colours.

Hedreen, Guy. 2017. "Unframing the Representation: The Frontal Face in Athenian Vase-Painting." In *The Frame in Classical Art: A Cultural History*, ed. V. Platt and M. Squire, 154–87. Cambridge: Cambridge University Press.

Hemingway, Colette, and Sean Hemingway. 2003. "The Technique of Bronze Statuary in Ancient Greece." *Heilbrunn Timeline of Art History*. New York: Metropolitan Museum.

Hersey, George. 2009. *Falling in Love with Statues: Artificial Humans from Pygmalion to the Present*. Chicago: University of Chicago Press.

Higley, Sarah L. 1997. "Alien Intellect and the Robotization of the Scientist." *Camera Obscura* 14, 1–2:131–60.

Hinuber, Oskar von. 2010. "Did Hellenistic Kings Send Letters to Aśoka?" *Journal of the American Oriental Society* 130, 2:261–66.

Hodges, Henry. 1970. *Technology in the Ancient World*. Harmondsworth: Penguin.

Hsing, I-Tien, and William Crowell. 2005. "Heracles in the East: The Diffusion and Transformation of His Image in the Arts of Central Asia, India, and Medieval China." *Asia Major*, 3rd ser., 18, 2:103–54.

Huffman, Carl. 2003. *Archytas of Tarentum: Pythagorean, Philosopher, and Mathematician King*. Cambridge: Cambridge University Press.

Ienca, Marcello, and Roberto Andorno. 2017. "Towards a New Human Rights in the Age of Neuroscience and Neurotechnology." *Life Sciences, Society and Policy* 13, 5. https://lsspjournal.springeropen.com/articles/10.1186/s40504-017-0050-1.

Iverson, Paul. 2017. "The Calendar on the Antikythera Mechanism and the Corinthian Family of Calendars." *Hesperia* 86, 1:129–203.

Jacobsen, Annie. 2015. "Engineering Humans for War." *Atlantic*, September 23. https://www.theatlantic.com/international/archive/2015/09/military-technology-pentagon-robots/406786/.

James, Peter, and Nick Thorpe. 1994. *Ancient Inventions*. New York: Ballantine.

Jansari, Sushma. 2011. "Buddhism and Diplomacy in Asoka's Embassies to the Mediterranean World." Lecture, Royal Asiatic Society of Great Britain and Ireland, London, December 14.

Kalligeropoulos, D., and S. Vasileiadou. 2008. "The Homeric Automata and Their Implementation." In *Science and Technology in Homeric Epics*, ed. S. A. Paipetis, 77–84. New York: Springer Science + Business Media.

Kang, Minsoo. 2005. "Building the Sex Machine: The Subversive Potential of the Female Robot." *Intertexts*, March 22.

———. 2011. *Sublime Dreams of Living Machines: The Automaton in the European Imagination*. Cambridge, MA: Harvard University Press.

Kantrowitz, Alex. 2017. "Microsoft's Chatbot Zo Calls the Qur'an Violent and Has Theories about Bin Laden." BuzzFeed News, July 3. www.buzzfeed.com/alexkantrowitz/microsofts-chatbot-zo-calls-the-quran-violent-and-has?utm_term=.mm7d6Rz1x#.ct8QlA7qj.

Kaplan, Matt. 2015. *Science of the Magical*. New York: Scribner.

Keats, Jonathon. 2017. "Caring Computers: Designing a Moral Machine." *Discover*, May. http://discovermagazine.com/2017/may-2017/caring-computers.

Keay, John. 2011. *India: A History, Revised and Updated*. New York: Grove Atlantic.

Keen, Antony. 2015. "SF's Rosy-Fingered Dawn." In Rogers and Stevens 2015, 105–20.

Keyser, Paul. 1993. "The Purpose of the Parthian Galvanic Cells: A First Century A.D. Electric Battery Used for Analgesia. *Journal of Near Eastern Studies* 52, 2:81–98.

———. 2016. "Venus and Mercury in the Grand Procession of Ptolemy II." *Historia* 65, 1:31–52.

Keyser, Paul, and Georgia Irby-Massie, eds. 2008. *The Encyclopedia of Ancient Natural Scientists*. London: Routledge.

Klein-Rogge, Rudolf. 1927. "The Creation of the Artificial Human Being." *Metropolis Magazine*, 32-page program for film premiere at Marble Arch Pavilion, London.

Knight, Will. 2017. "The Dark Secret at the Heart of AI." *MIT Technology Review*, April 11. https://www.technologyreview.com/s/604087/the-dark-secret-at-the-heart-of-ai/.

Koetsier, Teun, and Hanfried Kerle. 2015. "The Automaton Nysa: Mechanism Design in Alexandria in the 3rd Century BC." In *Essays on the History of Mechanical Engineering*, ed. F. Sorge and G. Genchi, 347–66. Cham, Switzerland: Springer.

Konstam, Nigel, and Herbert Hoffmann. 2004. "Casting the Riaci Bronzes: A Sculptor's Discovery." *Oxford Journal of Archaeology* 23, 4:397–402.

Korshak, Yvonne. 1987. *Frontal Faces in Attic (Greek) Vase Painting of the Archaic Period*. Chicago: Ares.

Kotsanas, Kostas. 2014. *Ancient Greek Technology: Inventions of the Ancient Greeks*. Katakalo, Greece: Kotsanas Museum of Ancient Greek Technology.

Krauskopf, Ingrid. 1995. *Heroen, Götter und Dämonen auf etruskischen Skarabäen. Listen zur Bestimmung*. Mannheim: University of Heidelberg.

Kris, Ernst, and Otto Kurz. 1979. *Legend, Myth, and Magic in the Image of the Artist*. New Haven, CT: Yale University Press.

Lachman, Gary. 2006. "Homunculi, Golems, and Artificial Life." *Quest* 94, 1 (January–February): 7–10.

LaGrandeur, Kevin. 2013. *Androids and Intelligent Networks in Early Modern Literature and Culture*. New York: Routledge.

Lane, George S. 1947. "The Tocharian Punyavantajataka, Text and Translation." *Journal of the American Oriental Society* 67, 1:33–53.

Lane Fox, Robin. 2009. *Travelling Heroes in the Epic Age of Homer*. New York: Knopf.

Lefkowitz, Mary. 2003. *Greek Gods, Human Lives: What We Can Learn from Myths*. New Haven, CT: Yale University Press.

Legge, James, trans. 1965 [1886]. *A Record of Buddhist Kingdoms . . . by the Chinese Monk Fa-Hien of Travels in India . . . (AD 399–414)*. . . . New York: Dover reprint.

Leroi, Armand Marie. 2014. *The Lagoon: How Aristotle Invented Science*. London: Bloomsbury.

Lichtheim, Miriam. 1980. *Ancient Egyptian Literature*. Vol. 3, *The Late Period*. Berkeley: University of California Press.

Lin, Patrick. 2012. "More Than Human? The Ethics of Biologically Enhancing Soldiers." *Atlantic*, February 16. http://www.theatlantic.com/technology/archive/2012/02/more-than-human-the-ethics-of-biologically-enhancing-soldiers/253217/.

———. 2015. "Do Killer Robots Violate Human Rights?" *Atlantic*, April 20. https://www.theatlantic.com/technology/archive/2015/04/do-killer-robots-violate-human-rights/390033/

Lin, Patrick, et al. 2014. "Super Soldiers (Part 1): What Is Military Human Enhancement? (Part 2) The Ethical, Legal and Operational Implications." In *Global Issues and Ethical Considerations in Human Enhancement Technologies*, ed. S. J. Thompson, 119–60. IGI Global.

Lin, Patrick, Keith Abney, and George Bekey, eds. 2014. *Robot Ethics: The Ethical and Social Implications of Robotics.* Cambridge, MA: MIT Press.

Lin, Patrick, Ryan Jenkins, and Keith Abney, eds. 2017. *Robot Ethics 2.0: From Autonomous Cars to Artificial Intelligence.* Oxford: Oxford University Press.

Liu, Lydia. 2011. *The Freudian Robot.* Chicago: University of Chicago Press.

"Longevity: Adding Ages." 2016. *Economist*, August 13, 14–16.

Lowe, Dunstan. 2016. "Suspending Disbelief: Magnetic and Miraculous Levitation from Antiquity to the Middle Ages." *Classical Antiquity* 35, 2:247–78.

Mackey, Douglas. 1984. "Science Fiction and Gnosticism." *Missouri Review* 7:112–20.

Maldonado, Alessandra. 2017. "This Man Had an Awkward Conversation with an A.I. Sex Robot So You Don't Have To." *Salon*, August 10. https://www.salon.com/2017/08/10/this-man-had-an-awkward-conversation-with-an-a-i-sex-robot-so-you-dont-have-to/.

Maluf, N.S.R. 1954. "History of Blood Transfusion: The Use of Blood from Antiquity through the Eighteenth Century." *Journal of the History of Medicine and Allied Sciences* 9:59–107.

Mansfield, Justin. 2015. "Models, Literary and Wax: The Fantastic in Demotic Tales and the Greek Novel." Paper delivered at the International Conference on the Fantastic in the Arts, Orlando FL, March 18–21.

Marconi, Clemente. 2009. "Early Greek Architectural Decoration in Function." In *Koine: Mediterranean Studies in Honor of R. Ross Holloway*, ed. D. Counts and A. Tuck. Oxford: Oxbow Books.

Marsden, E. W. 1971. *Greek and Roman Artillery.* Oxford: Oxford University Press.

Marshall, C. W. 2017. "Do Androids Dream of Electric Greeks?" *Eidolon*, October 26. https://eidolon.pub/do-androids-dream-of-electric-greeks-a407b583a364.

Martinho-Truswell, Antone. 2018. "To Automate Is Human." *Aeon*, February 13. https://aeon.co/essays/the-offloading-ape-the-human-is-the-beast-that-automates.

Martini, W. 1971. *Die etruskische Ringsteinglyptik.* Heidelberg: F. H. Kerle.

Matheson, Susan. 1995a. *Polygnotus and Vase Painting in Classical Athens.* Madison: University of Wisconsin Press.

———. 1995b. "The Mission of Triptolemus and the Politics of Athens." *Greek, Roman and Byzantine Studies* 35, 4:345–72.

Mattusch, Carol. 1975. "Pollux on Bronze Casting: A New Look at *kanabos*." *Greek, Roman and Byzantine Studies* 16, 3:309–16.

Mayor, Adrienne. 2007. "Mythic Bio-Techne in Classical Antiquity: Hope and Dread." Biotechnique Exhibit Catalogue. San Francisco: Yerba Buena Center for the Arts.

———. 2009. *Greek Fire, Poison Arrows, and Scorpion Bombs: Biological and Chemical Warfare in the Ancient World.* Rev. ed. New York: Overlook-Duckworth.

———. 2010. *The Poison King: The Life and Legend of Mithradates, Rome's Deadliest Enemy.* Princeton, NJ: Princeton University Press.

———. 2014. *The Amazons: Lives and Legends of Warrior Women across the Ancient World.* Princeton, NJ: Princeton University Press.

———. 2016. "Bio-Techne Myths: What Can the Ancient Greeks Teach Us about Artificial Intelligence, Robots, and the Quest for Immortality?" *Aeon*, May. https://aeon.co/essays/replicants-and-robots-what-can-the-ancient-greeks-teach-us.

McEvilley, Thomas. 2001. *The Shape of Ancient Thought: Comparative Studies in Greek and Indian Philosophies.* New York: Allworth.

McFadden, Robert. 1988. "Daedalus Flies from Myth into Reality." *New York Times*, April 24. http://www.nytimes.com/1988/04/24/world/daedalus-flies-from-myth-into-reality.html.

McKeown, J. C. 2013. *A Cabinet of Greek Curiosities.* Oxford: Oxford University Press.

Mendelsohn, Daniel. 2015. "The Robots Are Winning!" *New York Review of Books*, June 4.

Michaelis, Anthony. 1992. "The Golden Honeycomb: A Masterly Sculpture by Michael Ayrton." *Interdisciplinary Science Reviews* 17, 4:312.

Mill, John Stuart. 1859. *On Liberty.* London: Parker and Son.

Misailidou-Despotidou, Vasiliki. 2012. "A Red-Figure Lekythos by the Pan Painter from Ancient Aphytis." In *Threpteria: Studies on Ancient Macedonia*, ed. T. Michalis et al., 215–39. Thessaloniki.

Mori, Masahiro. 1981 [1974]. *The Buddha in the Robot: A Robot Engineer's Thoughts on Science and Religion.* Trans. Charles Terry. Tokyo: Kosei; originally published in 2 vols., 1974.

———. 2012 [1970]. "The Uncanny Valley." Trans. Karl F. MacDorman and Norri Kageki, authorized by Masahiro Mori. *IEEE Robotics & Automation Magazine*, 98–100; originally published in Japanese, in *Energy* 7, 4 (1970): 33–35. http://goo.gl/iskzXb.

Morris, Sarah. 1992. *Daidalos and the Origins of Greek Art.* Princeton, NJ: Princeton University Press.

Needham, Joseph. 1991. *Science and Civilization in China.* Vol. 2, *History of Scientific Thought.* Cambridge: Cambridge University Press.

Needham, Joseph., 1986. *Science and Civilisation: Mechanical Engineering.* Vol. 4, pt. 2. Taipei: Caves Books.

Needham, Joseph, and Ling Wang. 1965. *Science and Civilisation: Physics and Physical Technology, Mechanical Engineering.* Vol. 4, pt. 2. Cambridge: Cambridge University Press; abridged version by Conan Alistair Ronan. 1994.

Neils, Jenifer. 2005. "The Girl in the *Pithos*: Hesiod's *Elpis*." In *Periklean Athens and Its Legacy: Problems and Perspectives*, ed. J. Barringer and J. Hurwit, 37–46. Austin: University of Texas Press.

Newlands, Carole. 1997. "The Metamorphosis of Ovid's Medea." In Clauss and Johnston 1997, 178–209.

Newman, Heather. 2014. "The Talos Principle Asks You to Solve Puzzles, Ponder Humanity." *Venture Beat*, December 8. http://venturebeat.com/2014/12/08/the-talos-principle-asks-you-to-solve-puzzles-ponder-humanity-review/view-all/.

"The Next Frontier: When Thoughts Control Machines." 2018. *Economist, Technology Quarterly: Brain-Computer Interfaces, Thought Experiments*, January 6–12, 1–12.

Nissenbaum, Dion. 2014. "U.S. Military Turns to Hollywood to Outfit the Soldier of the Future." *Wall Street Journal*, July 4.

Nostrand, Anna van. 2015. "Ancient Bionics: The Origins of Modern Prosthetics." *Dig Ventures*, March 10. https://digventures.com/2015/03/ancient-bionics-the-origins-of-modern-prosthetics/.

Oleson, John Peter. 2009. *The Oxford Handbook of Engineering and Technology in the Classical World*. Oxford: Oxford University Press.

O'Sullivan, Patrick. 2000. "Satyr and Image in Aeschylus' *Theoroi*." *Classical Quarterly* 50:353–66.

Paipetis, S. A., ed. 2010. *Unknown Technology in Homer*. New York: Springer Science.

Palaephatus. *On Unbelievable Tales*. 1996. Trans. and comm. Jacob Stern. Wauconda, IL: Bolchazy-Carducci.

Pannikar, R. 1984. "The Destiny of Technological Civilization: An Ancient Buddhist Legend, Romavisaya." *Alternatives* 10 (Fall):237–53.

Panofsky, Dora, and Erwin Panofsky. 1991. *Pandora's Box: The Changing Aspects of a Mythical Symbol*. Princeton, NJ: Princeton University Press.

Paratico, Angelo. 2014. "Are the Giants of Mount Prama Odyssey's Laestrygonians?" *Beyond Thirty-Nine*, June 2. https://beyondthirtynine.com/are-the-giants-of-mount-prama-odysseys-laestrygonians/.

Pollitt, J. J. 1990. *The Art of Ancient Greece: Sources and Documents*. Cambridge: Cambridge University Press.

Pomeroy, Sarah. 2002. *Spartan Women*. Oxford: Oxford University Press.

Popcorn, Faith. 2016. "The Humanoid Condition." In *Economist* special issue *The World in 2016*, January.

Poulsen, Frederik. 1945. "Talking, Weeping and Bleeding Statues." *Acta Archaeologica* 16:178–95.

Raggio, Olga. 1958. "The Myth of Prometheus: Its Survival and Metamorphoses up to the Eighteenth Century." *Journal of the Warburg and Courtauld Institutes* 21, 1–2:44–62.

Raghavan, V. 1952. *Yantras or Mechanical Contrivances in Ancient India*. Bangalore: Indian Institute of Culture.

Raphael, Rebecca. 2015. "Disability as Rhetorical Trope in Classical Myth and *Blade Runner*." In Rogers and Stevens 2015, 176–96.

Raven, Maarten Jan. 1983. *Wax in Egyptian Magic and Symbolism*. Leiden: RMO.

Reeder, Ellen. 1995. *Pandora: Women in Classical Greece*. Baltimore: Walters Art Museum.

Reeve, C.D.C. 2017. "Sex and Death in Homer: Unveiling the Erotic Mysteries at the Heart of the Odyssey." *Aeon*, February 16. https://aeon.co/essays/unveiling-the-erotic-mysteries-at-the-heart-of-homers-odyssey.

Reid, Jane Davidson. 1993. *Classical Mythology in the Arts, 1300s-1990s*. 2 vols. Oxford: Oxford University Press.

Reynolds, Matt. 2017. "Peering Inside an AI's Brain Will Help Us Trust Its Decisions." *New Scientist*, July 3. www.newscientist.com/article/2139396-peering-inside-an-ais-brain-will-help-us-trust-its-decisions/.

Richardson, Arlan. 2013. "Rapamycin, Anti-Aging, and Avoiding the Fate of Tithonus." *Journal of Clinical Investigation* 123, 8 (August): 3204–6.

Richey, Jeffrey. 2011. "I, Robot: Self as Machine in the *Liezi*." In *Riding the Wind with Liezi: New Perspectives on a Daoist Classic*, ed. Ronnie Littlejohn, and Jeffrey Dippmann, 193–208. Albany: SUNY Press.

Richter, Gisela. 2006. *Catalogue of Engraved Gems: Greek, Roman and Etruscan*. Rome: L'Erma di Bretschneider.

Robertson, Martin. 1997. "The Death of Talos." *Journal of Hellenic Studies* 97:158–60.

Rogers, Brett, and Benjamin Stevens, eds. 2015. *Classical Traditions in Science Fiction*. Oxford: Oxford University Press.

Ruffell, Ian. 2015–16. "Hero's Automata: First Moves." "Riding the Snail." University of Glasgow, Classics, research blog on ancient technology. http://classics.academicblogs.co.uk/heros-automata-first-moves/; http://classics.academicblogs.co.uk/riding-the-snail/.

Sage, Evan T. 1935. "An Ancient Robotette." *Classical Journal* 30, 5:299–300.

Scarisbrick, Diana, Claudia Wagner, and John Boardman. 2016. *The Beverley Collection of Gems at Alnwick Castle*. Oxford: Classical Art Research Center.

Schmidt, Victor. 1995. *A Legend and Its Image: The Aerial Flight of Alexander the Great in Medieval Art*. Groningen: Egbert Forsten.

Scobie, Alex, and A.J.W. Taylor. 1975. "Perversions Ancient and Modern: I. Agalmatophilia, the Statue Syndrome." *Journal of the History of Behavioral Sciences* 11, 1:49–54.

Shapiro, H. A. 1994. *Myth into Art: Poet and Painter in Classical Greece*. London: Routledge.

Shelley, Mary. 1831 [1818]. *Frankenstein, or The Modern Prometheus*. Rev. ed. London: Colburn.

Shtulman, Andrew. 2017. *Scienceblind: Why Our Intuitive Theories about the World Are So Often Wrong*. New York: Basic Books.

Simons, G. L. 1992. *Robots: The Quest for Living Machines*. London: Cassell.

Sinclair, K. D., et al. 2016. "Healthy Ageing of Cloned Sheep." *Nature Communications* 7. https://www.nature.com/articles/ncomms12359.

Sladek, John. 1983. *Tik-Tok*. London: Corgi.

"The Sounding Statue of Memnon." 1850. *Fraser's Magazine for Town and Country* 42 (September): 267–78.

Spier, Jeffrey. 1992. *Ancient Gems and Finger Rings: Catalogue of the Collection*. Malibu, CA: J. Paul Getty Museum Publications.

Spivey, Nigel. 1995. "Bionic Statues." In *The Greek World*, ed. Anton Powell, 442–59. London: Routledge.

Steiner, Deborah. 2001. *Images in Mind: Statues in Archaic and Classical Greek Literature and Thought*. Princeton, NJ: Princeton University Press.

Stoneman, Richard. 2008. *Alexander the Great: A Life in Legend*. New Haven, CT: Yale University Press.

Stormorken, H. 1957. "Species Differences of Clotting Factors in Ox, Dog, Horse, and Man." *Acta Physiologica* 41:301–24.

Strong, John S. 2004. *Relics of the Buddha*. Princeton, NJ: Princeton University Press.

Summerville, Adam, et al. 2017. "Procedural Content Generation via Machine Learning (PCGML)." *arXiv preprint arXiv:1702.00539*, 1–15. https://arxiv.org/pdf/1702.00539.pdf.

Tanz, Jason. 2016. "The End of Code." *Wired*, June, 72–79.

Tassinari, Gabriella. 1992. "La raffigurazione de Prometeo creatore nella glittica romana." *Xenia Antiqua* 1:61–116.

———. 1996. "Un bassorilievo del Thorvaldsen: Minerva e Prometeo." *Analecta Romana Instituti Danici* 23:147–76.

Tenn, William. 1958. "There Are Robots among Us." *Popular Electronics*, December, 45–46.

Truitt, E. R. 2015a. *Medieval Robots: Mechanism, Magic, Nature, and Art*. Philadelphia: University of Pennsylvania Press.

———. 2015b. "Mysticism and Machines." *History Today* 65, 7 (July).

Tyagi, Arjun. 2018. "Augmented Soldier: Ethical, Social and Legal Perspective." *Indian Defence Review*, February 1. http://www.indiandefencereview.com/spotlights/augmented-soldier-ethical-social-legal-perspective/.

Vallianatos, Evaggelos. 2017. "The Shield of Achilles." *Huffington Post*, September 15.

Van Wees, Hans. 2013. "A Brief History of Tears." In *When Men Were Men*, ed. L. Foxhall and J. Salmon, 10–53. London: Routledge.

Voon, Claire. 2017. "The Sophisticated Design of a 3,000-Year-Old Wooden Toe." *Hyperallergic*. https://hyperallergic.com/387047/the-sophisticated-design-of-a-3000-year-old-wooden-toe/.

Vulpio, Carlo. 2012. "Il mistero dei giganti." *Corriere della Serra* (Milan), September.

Walker, Casey, ed. 2000. *Made Not Born: The Troubling World of Biotechnology*. San Francisco: Sierra Club Books.

Walton, Jo. 2015. *The Just City*. Bk. 1 of the trilogy *Thessaly*. New York: Tor Books.

Wang, Michelle. 2016. "Early Chinese Buddhist Sculptures as Animate Bodies and Living Presences." *Ars Orientalis* 46:13–38.

Weiner, Jesse. 2015. "Lucretius, Lucan, and Mary Shelley's *Frankenstein*." In Rogers and Stevens 2015, 46–74.

Weinryb, Ittai. 2016. *The Bronze Object in the Middle Ages*. Cambridge: Cambridge University Press.

West, Martin L. 2005. "The New Sappho." *Zeitschrift für Papyrologie und Epigraphik* 151:1–9.

Wilson, Emily. 2004. *Mocked with Death: Tragic Overliving from Sophocles to Milton.* Baltimore: Johns Hopkins University Press.

Winkler, Martin. 2007. "Greek Myth on the Screen." In *Cambridge Companion to Greek Mythology*, ed. Roger Woodard, 453–79. Cambridge: Cambridge University Press.

Wood, Gaby. 2002. *Edison's Eve: A Magical History of the Quest for Mechanical Life.* New York: Anchor Books.

Woodcroft, Bennett, ed. and trans. 1851. *The Pneumatics of Hero of Alexandria.* London: Taylor Walton and Maberly.

Woodford, Susan. 2003. *Images of Myths in Classical Antiquity.* Cambridge: Cambridge University Press.

Wosk, Julie. 2015. *My Fair Ladies: Female Robots, Androids, and Other Artificial Eves.* New Brunswick, NJ: Rutgers University Press.

Yan, Hong-Sen, and Marco Ceccarelli, eds. 2009. *International Symposium on History of Machines and Mechanisms.* New York: Springer Science + Business Media.

Zarkadakis, George. 2015. *In Our Own Image: Savior or Destroyer? The History and Future of Artificial Intelligence.* New York: Pegasus.

Zielinski, Siegfried, and Peter Weibel, eds. 2015. *Allah's Automata: Artifacts of the Arab-Islamic Renaissance (800–1200).* Karlsruhe: ZKM.

Zimmer, Carl. 2016. "What's the Longest a Person Can Live?" *New York Times*, October 5. https://www.nytimes.com/2016/10/06/science/maximum-life-span-study.html?_r=0.

INDEX

ILLUSTRATIONS ARE INDICATED BY PAGE NUMBERS IN ITALIC TYPE.